T0137472

Big Data Technologies for Monitoring
of Computer Security: A Case Study
of the Russian Federation

Sergei Petrenko

Big Data Technologies for Monitoring
of Computer Security: A Case
Study
of the Russian Federation

 Springer

Sergei Petrenko
Innopolis University
Innopolis, Tatarstan Republic, Russia

ISBN 978-3-030-07711-2   ISBN 978-3-319-79036-7   (eBook)
https://doi.org/10.1007/978-3-319-79036-7

Printed on acid-free paper

This Springer imprint is published by the registered company Springer International Publishing AG part of Springer Nature.
The registered company address is: Gewerbestrasse 11, 6330 Cham, Switzerland

Foreword: Alexander Tormasov

Dear readers!

This book shares valuable insight gained during the process of designing and constructing open segment prototypes of an early-warning cybersecurity system for critical national infrastructure in the Russian Federation. In preparing its publication, great attention was given to the recommendations and requirements set out in *the concept of state systems for detecting, preventing, and eliminating the consequences of cyber-attacks on information resources of the Russian Federation (approved by the President of the Russian Federation on December, 12, 2014, Ns K 1274), as well as best international practices* that have been gained in this field.

According to data provided by the Innopolis University Information Security Center, the number of computer attacks is continuously rising, with only 45% of them officially registered and 55% remaining undetected and thus unprevented.

The modern level of development in information and communication technologies (ICT) now makes it possible to take industrial production and scientific research in information security to a fundamentally higher plane, but the effectiveness of such a transition directly depends on the availability of highly qualified specialists. Every year, about 5000 Russian specialists graduate in the field of information security, whereas the actual industrial demand is estimated at *21,000 per year* through 2020. For this reason, the Russian Ministry of Education and Science, along with executive governmental bodies, has created a high-level training program, which they continually develop, for state information security employees. This initiative includes *170 universities, 40 institutions of continuing education, and 50 schools of secondary vocational training.* In evaluating the universities' performance over *30 academic disciplines, information security has scored the highest for three consecutive years on the Russian Unified State Examination (Единый Государственный Экзамен). In addition, employee training subsystems operating in the framework of the Russian Federal Security Service, the Russian Ministry of Defense, the Russian Federal Protective Service, Russian Federal Service for Technical and Export Control, and the Russian Emergencies Ministry of Emergency Situations are similar to the general system for training information security specialists at the Russian Ministry of Education and Science, which trains personnel according to the concrete needs of individual departments.*

Yet, there remains the well-known *problem* that the vast majority of educational programs in information security struggle to keep pace with the rapid development in the ICT sphere, where significant changes occur every 6 months. As a result, existing curricula and programs do not properly train graduates for the practical reality of what it means to efficiently solve modern information security problems. For this reason, graduates often find themselves lacking the actual skills in demand on the job market. In order to ensure that education in this field truly satisfies modern industrial demands, Innopolis University students and course participants complete actual information security tasks for commercial companies as well as governmental bodies (e.g., for the university's over 100 industrial partners). Also, Innopolis University students participate in domestic and international computer security competitions, e.g., the game *Capture the Flag* (CTF), considered to be among the most authoritative in the world.

Currently, Innopolis University trains information security specialists in "Computer Science and Engineering" (MA program in Secure Systems and Network Design). The program is based on the University of Amsterdam's "System and Network Engineering" program with its focus on information security. In 2013, it was ranked as the best MA program for IT in the Netherlands (Keuzegids Masters 2013), and in 2015 it won the award for best educational program (Keuzegids Masters 2015). The University of Amsterdam is one of Innopolis University's partners and is included in the Top 50 universities of the world (QS World university rankings, 2014/2015). An essential feature of this program is that Innopolis University students take part in relevant research and scientific-technical projects from the beginning of their studies. In solving computer security tasks, students have access to the scientific-technical potential of 3 institutes, 13 research laboratories, and 3 research centers engaged in advanced IT research and development at Innopolis University. This partnership also extends to Innopolis University's academic faculty, both pedagogic and research-oriented, which numbers more than 100 world-class specialists.

The information security education at Innopolis University meets the core curriculum requirements set out in the *State Educational Standards for Higher Professional Education 075 5000 "Information Security" in the following degrees: "Computer Security," "Organization and Technology of Information Security," "Complex Software Security," "Complex Information Security of Automated Systems," and "Information Security of Telecommunication Systems."* At the same time, high priority is given to practical security issues of high industrial relevance; however, given the relative novelty of these needs, they remain insufficiently addressed in the curricula of most Russian universities and programs. These issues include the following:

- Computer Emergency Response Team (CERT) based on groundbreaking cognitive technologies
- Trusted cognitive supercomputer and ultra-high performance technologies

- Adaptive security architecture technologies
- Intelligent technologies for ensuring information security based on big data and stream processing (BigData + ETL)
- Trusted device mesh technology and advanced system architecture
- Software-defined networks technology (SDN) and network functions virtualization (NFV)
- Hardware security module technology (HSM)
- Trusted "cloud" and "foggy" computing, virtual domains
- Secure mobile technologies of 4G +, 5G, and 6G generations
- Organization and delivery of national and international cyber-training sessions
- Technologies for automated situation and opponent behavior modeling (WarGaming)
- Technologies for dynamic analysis of program code and analytical verification
- Quantum technologies for data transmission, etc.

The current edition of the *Big Data Technologies for Monitoring of Computer Security: A Case Study of the Russian Federation was written by Sergei Petrenko, Prof. Dr. Ing., Head of the Information Security Center at Innopolis University and Alexey Petrenko,* author and coauthor of more than 40 articles on information security issues. The work of these authors has significantly contributed to the creation of a national training system for highly qualified employees in the field of computer and data security technologies. This book sets out a notion of responsibility in training highly qualified specialists at the international level and in establishing a solid scientific foundation, which is prerequisite for any effective application of information security technologies.

Rector of the Innopolis University,   Alexander Tormasov
Innopolis, Russia

Foreword: Igor Kalyaev

Nowadays, the information confrontation plays an increasingly important role in modern, "hybrid" wars. Furthermore, victory is often attained not only via military or numerical superiority, but rather by information influence on various social groups or by cyber-attacks on critically important governmental infrastructure.

In this regard, means for detecting and preventing information and technical impacts should play a crucial role. Currently, systematic work is being done in Russia to create a National Cyber-attack Early-Warning System. A number of state and corporate cybersecurity response system centers have already been organized. However, the technologies applied in these centers allow only the detection and partial reflection of ongoing IT-attacks, but they do not have the capacity to predict and prevent attacks that are still in the preparation stage.

Such a situation requires the creation of fundamentally new information security systems, which are capable of controlling the information space, generating and simulating scenarios for the development, prevention, and deterrence of destructive information and technical impacts, and of initiating proactive responses to minimize their negative impact. New technologies in big data and deep learning as well as in semantic and cognitive analysis are now capable of proactively identifying the invader's hidden meanings and goals, which the other types of analysis could not discover, will likely play an instrumental role here. This monograph aims to develop these methods and technologies.

At the same time, it is impossible to implement a National Cyber-attack Early-Warning System without also tackling a series of related issues. Most notably, this will necessarily entail the creation of an effective computing infrastructure that provides the implementation of new methods and technologies for modeling the development, prevention, and deterrence of destructive information and technical impacts in real-time, or even preemptively. Clearly, this problem will not be solved without high-performance computing systems or a supercomputer.

We must confess that Russia currently lags far behind leading Western countries in terms of its supercomputer technology. Cluster supercomputers primarily used in our country are usually based on a CKD assembly from commercially available foreign processing nodes and network switches. It is well known that this class of supercomputers demonstrates its optimal performance when solving loosely bound problems not requiring intensive data exchange between processor nodes. The actual performance of cluster supercomputers, however, is significantly reduced when solving tightly bound problems, in particular semantic and cognitive analysis of big data. Moreover, the attempts to increase the cluster system performance by increasing the number of processing nodes have often not only failed to yield positive results but, on the contrary, have had the opposite effect due to a heightened proportion of nonproductive "overhead" in the total solution time which arises not from "useful" processing, but from organizing a parallel calculation process. These fundamental disadvantages of modern cluster supercomputers are a product of their "hard" architecture, which is implemented at the stage of computer construction and cannot be modified while being used.

Developed by Russian scientists, the concept of creating a reconfigurable super-computer made it possible to configure the architecture setup (adjustment) depending on the structure of the task's solution without entailing the aforementioned disadvantages. In this case, a set of field programmable logic devices (FPLG) of a large integration degree comprises the entire computing field and enables the user to create the task-oriented computing structures similar to the graph algorithm of the given task; this is used as a supercomputer computational device, rather than a standard microprocessor. This approach ensures a "granulated" parallel computing process as well as a high degree of time efficiency in organization achieved by adjusting the computing architecture to the applied task. As a result, near-peak performance of the computing system is achieved and its linear growth is provided, when the hardware resources of the FPLG computational field are increased.

Today, reconfigurable FPLG-based computing systems are increasingly finding use in solving a number of topical applied tasks, primarily computationally labor-intensive and "tightly coupled" streaming tasks that require mass data processing (streams), as well as tasks that require the processing of nonstandard data formats or variable number of bit (e.g., applied fields of big data semantic and cognitive analysis, cryptography, images processing and recognition, etc.). This allows us to estimate the prospects of using reconfigurable supercomputers technology when establishing a National Cyber-attack Early-Warning System.

At the same time, one supercomputer, even the most productive one, is not enough to create the computing infrastructure of the National Cyber-attack Early-Warning System. Obviously, such a system should be built based on a network of supercomputer centers, with each unit having its own task focus, while preserving the possibility to combine all the units into a single computing resource; this would, de facto, provide a solution to computationally labor-intensive tasks of real-time and preemptive modeling development scenarios for prevention and deterrence of the destructive information and technical impacts. In other words, the National Cyber-attack Early-Warning System should be based on a certain segment (possibly secured from outside users) of the National Supercomputer GRID network.

Furthermore, establishing a National Supercomputer GRID Network evokes a complex problem of optimal distribution (dispatching) of computational resources while solving a stream of tasks on modeling development scenarios for cyber-attack prevention and deterrence.

Nowadays, the problem of dispatching distributed computer networks is being solved with uniquely allocated server nodes. However, such centralized dispatching is effective when working with a small computational capacity or nearly homogenous computational resources. However, in cases of numerous, heterogenous network resources, the operational distribution (also redistribution) of tasks, not to mention informationally relevant subtasks via a single central dispatcher, becomes difficult to implement. Moreover, using a centralized dispatcher significantly reduces the reliability and fault tolerance of the GRID network, since a failure on the part of the service server node that implements the dispatcher functions will lead to disastrous consequences for the entire network.

These disadvantages can be avoided by using the principles of decentralized multiagent resource management of the GRID network. In this case, software agents that are physically implemented in each computational resource as part of the GRID network play the main role in the dispatching process and represent their "interests" in the dispatching process. Each agent will "know" the computing capabilities of "its own" resource, as well as responsively track all changes (e.g., performance degradation owing to the failure of numerous computing nodes). Given this information, the agent can "allocate" its resource for solving tasks where "its" resource will prove most effective. If the computing resource of one agent is not enough to solve the problem in the given time duration, then a community of agents will be created, with each one providing its resources for solving the various parts of a single task.

The benefits of a decentralized multiagent dispatching system in a National Supercomputer GRID network are manifold:

- Ensure efficient loading of all computational resources included in the GRID network, by using up-to-date information about their current status and task focus
- Ensure the adaptation of the computational process to all resource changes in the cloud environment
- Reduce the overhead costs for GRID network organization due to the absence of the need to include special service servers as a central dispatcher
- Increase the reliability and fault tolerance of the GRID network and, as a result, dependable computing, since the system will not have any elements whose failure may lead to disastrous consequences for the entire network.

The aforementioned problems are partially covered in this book; however, at the same time, they require further and deeper development.

In general, I believe that this monograph devoted to the solution of the urgent scientific and technical problem on the creation of the National Cyber-attack Early-Warning System is very useful for information security students, graduate students, scientists, and engineers specializing in the theory and practice of detecting, preventing, and deterring computer threats.

Member of Russian Academy of   Igor Kalyaev
Science, Southern Federal University,
Rostov-on-Don, Russia

Abstract

This scientific monograph considers possible solutions to the relatively new scientific-technical problem of developing an *early-warning cybersecurity system for critically important governmental information assets*. The solutions proposed are based on the results of exploratory studies conducted by the authors in the areas of big data acquisition, cognitive information technologies (cogno-technologies), and "computational cognitivism," involving a number of existing models and methods.

The results obtained permitted the design of an early-warning cybersecurity system.

In addition, prototypes were developed and tested for software and hardware complexes of stream preprocessing and processing as well as *big data* storage security, which surpass the well-known solutions based on Cassandra and HBase in terms of performance characteristics.

As such, it became possible, for the first time ever, to synthesize scenarios of an *early-warning cybersecurity system* in cyberspace on extra-large volumes of structured and unstructured data from a variety of sources: Internet/Intranet and IoT/IIoT (*Big Data and Big Data Analytics*).

The book is designed for undergraduate and postgraduate students, for engineers in related fields, as well as for managers of corporate and state structures, chief information officers (CIO), chief information security officers (CISO), architects, and research engineers in the field of information security.

Introduction

This monograph owes its relevance to the necessity to resolve the contradiction between the increasing need to ensure information security for *critically important governmental information assets* amid growing security threats and the insufficiency of existing models, methods, and means of detecting, preventing, and neutralizing the consequences of cyber-attacks. Concretely, this scientific-technical problem concerns the development of an *early-warning system for cyber-attacks*, and resolving this problem entails the search for possible solutions to a number of new scientific-technical problems:

- Input data classification and identification of primary and secondary signs of cyber-attacks based on big data, big data systems, and *Internet/Intranet* and *IoT/ IIoT* networks
- Formation, storage, and processing of relevant patterns of early detection based on *Big Data + ETL*
- Multifactor forecasting of computer attacks on extremely large volumes of structured and unstructured information (*Big Data and Big Data Analytics*)
- Generation of new knowledge on the quantitative patterns of information confrontation in cyberspace
- Synthesis of optimal deterrence scenarios as well as training in early detection system, etc.

Russia has already established a number of state and corporate computer incident response centers. In terms of their functionality, these centers are similar to the foreign *CERT (Computer Emergency Response Team)*, *CSIRT (Computer Security Incident Response Team)*, *MSSP (Managed Security Service Provider)*, *MDR (Managed Detection and Response Services)*, *and SOC (Security Operations Center)*, among others.

These Russian centers are known as *Information Security Monitoring Centers* based on the system of the distributed situational centers (*SRSC*), Information security centers of the distributed situational centers system in Russian Federation state authorities, State and corporate segments of Monitoring in the Detection,

Prevention and Cyber Security Incident Response (*SOPCA*), Computer Attack Detection and Prevention System (*SPOCA*) of the Russian Ministry of Defense, Crisis Management Center (*CMC*) of Rosatom State Corporation, Information Security Monitoring Center and FinCERT Bank of Russia, CERT Rostec, System of traffic analysis and network attack detection (*SATOSA*), OJSC Rostelecom, the Information Security Threat Monitoring Center Gazprom, the Information Security Situation Center of the GPB Bank, Solar Security Joint Special Operations Command (*JSOC*) and Security Operation Center (*SOC +*), Kaspersky Lab ICS-CERT and the Anti Targeted Attacks Security Operation Center (*SOC*) of Kaspersky Lab, etc.

However, the operating experience from the abovementioned centers has shown that existing methods and means are insufficient for detecting and preventing impact to information and technical resources. The ability to accumulate, aggregate, and analyze masses of relevant information does not provide decision-makers with warning of terrorist attacks (being planned or conducted), mass Distributed Denial of Service attacks (*DDOS*), and Advanced Persistent Threat attacks (*APT*) on critical infrastructure. Instead, these situation centers are merely able to detect and partly reflect existing impacts to information-technical resources, but are not able to prevent and prohibit aggressive action in advance. Even the sum of all available technical means for detection, prevention, and neutralization of the consequences of cyber-attacks would not be able to anticipate the next attack or malicious activity, without appropriate modification and significant intervention from qualified information security experts.

This increasingly suggests that these issues would be best resolved via assistance from intelligent information systems capable of generating the specifications and scenarios for proactive behavior when confronted with destructive information-technical impact in cyberspace conflicts. For this reason, the established concept of building computer incident response centers based on *data management* technology, which can merely generate automated incident overviews and assess the data on the basis of preprogrammed scenarios, is being replaced by the new concept of *knowledge management* for dealing with both actual and presumed cyberspace warfare. Its distinguishing characteristic lies in its ability to create semantic and cognitive information-analytical systems as well as conduct automated real-time "intent analysis" and generate appropriate warning and deterrence scenarios (i.e., identify and leverage aspects of the opponents' intentions and purposes which remained hidden under other means of analysis). Thus, harnessing this new technology to create detection and prevention systems in corporate and state structures offers a feasible approach to the real challenges of modern-day cybersecurity.

It should be noted that the similar technologies have already come partially into use. For example, software solutions of *Palantir Technologies, Inc.* (USA) are widely used for data content analysis for the special forces, police, and US Department of Defense. *Palantir* acts as a provider of "5th layer" solutions, which analyze the interrelations among internal and external control subjects, and is considered to be one of the technological leaders in perspective situational centers development, along with *IBM*, *HP* and *SAP*, *RSA*, *Centrifuge*, *Gotham*, *i2*, *SynerScope*, *SAS Institute*, *Securonix*, *Recorded Future*, etc. These solutions center on visualization of *Big Data* from heterogeneous sources, which identifies synergy, connections, and anomalies among the objects and surrounding events (i.e., *Data Mining* with an emphasis on interactive visual analysis for the purpose of intelligence enhancement). Data is gleaned from various open and closed databases, structured and unstructured sources of information, social networks, media, and messengers. For instance, the *Gotham* system implements an original technology for generating and managing domain ontologies that conceptually generalizes heterogeneous data from multiple sources and arranges it meaningfully for effective teamwork and machine learning.

The term *governance of global cyberspace* was first mentioned in the *National Strategy for Homeland Security (Office of Homeland Security, 2002)*. Further, the term was developed by the *US Department of Homeland Security* in a number of government regulations in the context of information systems and electronic data protection, as well as "creating the conditions for achieving national cybersecurity goals." For instance, the *National Strategy for Combating Terrorism (GPO, 2003)*, the *National Strategy for Secure Cyberspace (GPO)*, and the *National Strategy for the Physical Protection of Critical Infrastructures and Key Assets (GPO, 2003)* clearly indicated the need to create a unified *National Cyberspace Security Response System (NCSRS)*. This system should include the relevant departmental and corporate centers for an Information Sharing and Analysis Center (*ISAC*).

In 2003, the *Department of Homeland Security established the Cybersecurity and Telecommunications Regulatory Authority*, which includes the *National Cyber Security Division (NCSD)* and the *United States Computer Emergency Readiness Team (US-CERT)*. *NCSD* was appointed to be responsible for the general coordination of the interagency cybersecurity collaboration, as well as for achieving international cooperation and interacting with representatives from the private sector. The US-CERT team, along with the respective center, assumed responsibility for the technical issues of detection and warning, prevention, and elimination of the consequences of cyber-attacks by emergency recovery of the US critical infrastructure.

In January 2008, the US President's Directive *"Comprehensive National Cybersecurity Initiative" (CNCI)* was approved and about USD 30 billion was allocated for the relevant research programs. However, in mid-2008 the Department of Homeland Security initiatives received a harsh critique; more precisely, it was stated that *US-CERT* "is not capable of conducting high-quality monitoring of threats to the security critical infrastructure and has limited capabilities to eliminate the consequences of cyber-attacks and cannot create a cyber analysis and warning system (*DHS Faces Challenges in the Establishment of Comprehensive National Capability, US Government Accountability Office Report, GAO-08-588, 2008*).

Some of the main reasons cited for distrust in the Department of Homeland Security initiatives include a shortage of qualified *US-CERT* employees and limited technical capabilities of the first cyber-attack prevention system *Einstein-I* (2003) (currently in service with *Einstein II* (2007) and *Einstein III* (2014) – respectively).

The *Report of the CSIS Commission on Cybersecurity for the 44th Presidency*[1] recommended taking the following actions:

- Raise the priority level of US critical infrastructure cybersecurity to an executive level (i.e., White House) status, as the Commission found the IMB's initiatives and efforts to be insufficient
- Develop a *national cybersecurity strategy* that clearly outlines the key improvements, purposes, and development priorities in this area[2]
- Develop national and international legal norms to ensure an appropriate cybersecurity level and improve the law enforcement system by appropriately expanding its jurisdiction in cyberspace
- Charge a government structure with the practical implementation of the national cybersecurity strategy (according to the commission, the Ministry of Defense, and other agencies in the US intelligence community possess the capacity and resources necessary to address the problem)
- Establish a national operating center to provide cybersecurity control with a focus on practically implementing activities, rather than on further planning in this area
- Organize a sensitization campaign explaining the relevance and importance of the national critical infrastructure cybersecurity issues Prepare and implement appropriate training and development programs for public and private sector employees
- Develop the mechanisms of interaction at the international level for developing the capacity for joint defensive and offensive actions in cyberspace and generally increase the security of national critical infrastructure
- Develop effective mechanisms for interaction between public and private sectors for qualitative cybersecurity research
- Increase the level of scientific-technical interaction with private-sector representatives
- Replicate the results of successful research and development work carried out for the public-sector customer on other economic sectors

Nowadays, almost all types of the US Armed Forces pay special attention to the issue of conducting cyberspace operations. Moreover, the *Air Force*, the *Navy*, and the *ground forces* of the US Army each carried out relatively independent studies of the military-technical issues relating to conducting information operations in cyberspace, organized the appropriate staffing measures, and determined the required human resources.

---

[1]*Report of the CSIS Commission on Cybersecurity for the 44th Presidency, Center for Strategic and International Studies. Washington D.C., 2008.)*

[2]*National Cybersecurity Strategy. Key Improvements Are Needed to Strengthen the Nation's Posture. Statement of David Powner. United State Government Accountability Office. GAO-09-432T Washington D.C., 2009*

In December 2006, the Joint Chiefs of Staff committee prepared a document entitled *"The National Military Strategy for Cyberspace Operations[3],"* which set out the following priorities for cyber operations:

– Obtaining and maintaining the initiative via integrated defensive and offensive operations in cyberspace
– Inclusion of cyber operations in the military planning system
– Development of the most effective forms and methods of conducting cyber operations
– Assessing the effectiveness of said cyber operations
– Development of cooperative programs between the Ministry of Defense and NATO partners, other US government agencies, as well as representatives of the defense industry complex
– Establishment of ongoing training programs and professional development system for Department of Defense (DOD) cybersecurity specialists
– Conducting the necessary organizational and staffing reorganizations
– Creation of the appropriate infrastructure

Initially, the US Air Force bore the responsibility for developing the methods of conducting cyber operations. In 2005, Air Force Commander M. Wynne stated that "the operations in cyberspace correlate with the traditional tasks of the U.S. Air Force, and now they will fly not only in the air and space, but also in cyberspace" (Victory in Cyberspace. An Air Force Association Special Report. 2007).

However, a number of high-ranking DOD officials did not share this opinion. In particular, the Chairman of the Combined Chiefs of Staff, Admiral M. Mullen, believed that cyberspace operations should be handled by the US Network Operations Command Center, which in 2008 was transformed into the US Navy Cyber Power. At that time, the US Navy Cyber Power was the leading military unit for conducting cyber operations.

This command was reinforced by the units of electronic intelligence and cryptographical security, as well as by the US Naval Space Command assets.[4]

The so-called 7th signal command – the first unit of the US Army – responsible for information security control of computer systems and networks was formed in 2009. At the same time, work began on the revision of documents regulating information operations by ground forces[5] and the combined forces[6] in order to gain further authority in cyberspace.

---

[3]*The National Military Strategy for Cyberspace Operations. Chairman of the Joint Chiefs of Staff. Washington, 2006*

[4]*Information Operations Primer. Fundamentals of Information Operations. Washington: US Army War College, 2008.*

[5]*Field Service Regulations, FM 3–13*

[6]*Jont Publications 3–13*

The US Army Concept of Operations for 2010–2024[7] set out the following directives for cyber operations:

- Detection – passive or active monitoring of the information and electromagnetic sphere to identify threats to information resources and data communication channels
- Interruption of the invader's access to information resources – awareness limitation in combat conditions and information resources protection (at the levels of hardware and software) from possible use or influence from invaders (i.e., antivirus, firewall, immunity to interference, electromagnetic pulse interference, etc.)
- Degradation and reduction of the invader's information potential – interference in the operability of the information technology equipment in order to reduce its combat stability and controllability (electronic suppression, network computer attacks, etc.)
- Destruction – a guaranteed destruction of the invader's electronic equipment using directed energy weapons or traditional kinetic warfare
- Monitoring and analysis – data collection on the condition of cybernetic and electromagnetic media with a mind to offensive and defensive cybernetic operations
- Response – defensive (reducing the effectiveness of invader's operations) and offensive (counter-punching) response
- Influence – distortion of the information perception by people or public institutions, as well as distortion of information circulation in machine and combined systems (machine-human, human-machine) for reorientation of their actions own purposes, for personal needs, etc.

Such an admitted lack of coordination among military units led US military leadership to concentrate their coordinating functions within a single structure – the National Security Agency (NSA).

In early spring 2009, US Secretary of Defense R. Gates signed an order to coordinate all cyberspace operations within the Joint Functional Component Command for Network Warfare (*JFCCNW*).

JFCCNW subordinated the Joint Tactical Force for Global Network Operations (*JTF-GNO*), under the supervision of Chief of the Defense Information Systems Agency (DISA), Major-General of the Ground Force, K. Pollet.

In fall 2009, the creation of United Cyber Command was announced under the supervision of Lieutenant-General K. Alexander, head of the NSA. The United Cyber Command was directly subordinate to the US Strategic Command and located at the Fort Meade military base in Maryland.

---

[7]*The United States Army Concept of Operations (CONOPS) for Cyber-Electronics (CE) 2010–2024. Concepts Development Division Capability Development Integration Directorate US Army Combined Arms Center: Author's Draft. 2009.*

In October 2010, a new cyber command was formed in the USA, with a motto of "second to none." This new unit, which combined preexisting cyberunits from the Pentagon (with approximately 21,000 staff members) was overseen first by Lieutenant-General K. Alexander and then by Admiral M. Rogers from April 2014 until present.

The tasks of the new *Joint Cyber Command* included the planning, coordination, integration, synchronization, and management of network operations and army network security. At the same time, the functional responsibilities of these cyberunits have been expanded to include cybersecurity control not only of military and state infrastructure but also of critical US commercial facilities.

Currently, the NSA manages a full range of issues on cyberspace control (including offensive operations, measures on protection of critical information infrastructure and information and telecommunications technologies) within the Department of Defense and at the national level. This seems reasonable, especially given the considerable amount of relevant experience in the agency. The ensuing redistribution of responsibilities greatly favored the NSA and highly prioritized prospective programs for the creation of a *High Assurance Platform* (*HAP*) and the development of a *Global Information Grid* (*GIG*).

The development of cyberspace information warfare programs in the USA has two main objectives.

Firstly, the development of prospective means to influence the information and telecommunication systems of a real and potential invader, including means of intercept control over unmanned aerial vehicles, disabling avionics, and other information equipment used in military systems, which veritably implies the discussion of a fundamentally new class of weapons – cyber weapons.

Secondly, implementing a program to create a highly protected computing architecture that will form the conditions for solidifying US superiority in the information and telecommunications sphere and provide support for American high-tech companies through direct government funding.

On this basis, the conclusion seems warranted that the scientific problem under discussion, the development of a scientific and methodical apparatus for giving early warning of cyber-attacks, has theoretical, scientific, and practical significance for all technologically developed states.

For instance, the urgency of creating an early detection system for a cyber-attack in the Russian Federation is confirmed by the requirements of the following legal documents.

This monograph is possibly the first to address the ongoing scientific-technical problem of developing an early detection system for a cyber-attack on a state's information resources. As such, every effort is made to consistently highlight the general motifs of the historical and current approaches and, thus, to do justice to the cognitive innovation in a consistent and coherent manner.

In this way, it becomes possible to independently associate and synthesize new knowledge concerning the qualitative characteristics and quantitative patterns of information confrontation.

This monograph proposes a "stage-by-stage" solution to the given scientific-technical problem.

Stage 1 – Design and development of a technical (structural) component of traditional detection, prevention, and elimination system for consequences of cyber-attacks based on big data technologies – creating a high-performance corporate (departmental) segment for work with big data.

Stage 2 – Creation of an analytical (functional) component based on the proposed methods of "computational cognitivism" – implementing the cognitive component of the system itself, capable of independently extracting and generating useful knowledge from large volumes of structured and unstructured information.

- The individual functions of this component will be handled in greater detail throughout the text.

This monograph is intended for the following reader groups:

- *Corporate and State CEO*, responsible for the proper information security provision and compliance with the relevant government requirements
- *Chief information officers* (*CIO*) and *Chief information security officers* (*CISO*), responsible for corporate information security programs and organization of the information security regime
- Database architect and research engineers responsible for the technical design of the *Security Threat Monitoring Centers* in the various *Situation Centers* and government (and corporate) segments of detection systems for the prevention of cyber-attacks

This book can also be a useful training resource for undergraduate and postgraduate students in related technical fields, since these materials are largely based on the authors' teaching experience at the *Moscow Institute of Physics and Technology* (*MIFT*) and *Saint Petersburg Electrotechnical University "LETI" n.a. V.I. Ulyanov* (*Lenin*)

This monograph contains four chapters devoted to the following subjects:

- The relevance of the given scientific-technical problem
- Establishing the finite capabilities of existing technologies for detecting and preventing cyber-attacks
- Limiting capabilities of the existing computing architectures of the von Neumann architecture determination
- Search of possible scientific-technical solutions to the problem of giving early warning of cyber-attacks on critical state infrastructure

The first chapter shows that the task of critical infrastructure security control is one of the most important tasks of *digital sovereignty* and *state defense capability*. The main threats to state information security, including threats of military-political, terrorist, and criminal nature, are demonstrated. Also, justification is given for the necessity of an integrated approach to ensure information security, not only at the national but also at the foreign policy level. Moreover, various concepts for ensuring information security without involving the military and political dimensions are shown to be ineffective. Examples of possible scenarios and technical methods of cyber-attacks on critical state infrastructure are considered. In sum, the problem of detecting and preventing cyberattacks is assessed as it currently stands.

The second chapter demonstrates the need to strengthen information security measures as a consideration of national security by heightening the level of state cyberspace control. Assessment is made of the limited technological capacity for detecting and preventing cyber-attacks. Similarly, appraisal is given of various corporate centers for monitoring information security threats to critical state infrastructure (CERT/SCIRT/MSSP/MDR/SOC). Furthermore, aspects of creating a "cloud" national response center for computer security incidents are discussed. This chapter aims to justify the need for a similar early-warning system on the basis of prospective information technologies.

The third chapter presents a plausible typification of evolutionary modifications for a "von Neumann architecture" for selecting a prospective hardware platform for a national cyber-attack early-warning system. This chapter also provides the program trajectory through 2025 for finding a solution on the basis of supercomputer technologies to the problem of developing an early-warning system.

The fourth chapter proposes an approach for creating an early-warning system based on "computational cognitivism": a relatively new field in scientific research where cognition and cognitive processes are a kind of *symbolic computation*. The cognitive approach permits the creation of systems, which fundamentally differ from traditional threat monitoring systems due to their unique ability to independently associate and synthesize new knowledge about qualitative characteristics and quantitative patterns of cyberspace information confrontations. In conclusion, this chapter proposes a possible early-warning system architecture based on the analysis and processing of extremely large amounts of structured and unstructured data from various Internet/Intranet and IoT/IIoT sources (*Big Data and Big Data Analytics*).

The book is written by leading research engineers of technical issues in information security, Doctor of Technical Sciences, prof. S.A. Petrenko, and research engineer A.S. Petrenko.

In advance, the authors would like to thank and acknowledge all readers. Anyone wishing to provide feedback or commentary may address the authors directly at:
s.petrenko@rambler.ru *and* a.petrenko1999@rambler.ru.

Russia-Germany   Sergei Petrenko
January 2018

Contents

Chapter 1
The Relevance of the Early Warning of Cyber-attacks

It is proved that the problem of information security of the critical infrastructure of the Russian Federation is one of the most important goals of ensuring digital sovereignty and defense capability of the state. The main threats to the information security of the Russian Federation are introduced. They include threats of military-political, terrorist, and criminogenic nature. The necessity of an integrated approach to information security not only at the national but also at the external policy level is explained. The current state of the problem of detection and prevention of cyber-attacks is assessed. Prospective assignments of alerting and anticipation tasks, as well as timely detection and neutralization of cyber-attacks, are considered.

1.1 The Modern Cyberthreat Landscape

On December 5, 2016, Russian President Vladimir Putin signed the Decree No. 646 on the approval of the new Information Security Doctrine of the Russian Federation, which develops the general provisions of the current concept of the Russian Federation's foreign policy in the field of information security.[1] The approved Doctrine is published on the official Internet portal of legal information, the state system of legal information.[2] Decree No. 646 came into force from the signing date, and the previous Information Security Doctrine of the Russian Federation, approved by the President of the Russian Federation on September 9, 2000 No. Pr-1895, was declared invalid. The

---

[1] *http://publication.pravo.gov.ru/Document/View/0001201612010045/*

[2] *http://publication.pravo.gov.ru/Document/View/0001201612060002?index=0&rangeSize=1/*

© Springer International Publishing AG, part of Springer Nature 2018
S. Petrenko, *Big Data Technologies for Monitoring of Computer Security: A Case Study of the Russian Federation*, https://doi.org/10.1007/978-3-319-79036-7_1

Doctrine defines strategic goals and basic directions of information security taking into account the strategic national priorities of the Russian Federation [1].

1.1.1   Modern World and Foreign Policy of the Russian Federation

The Foreign Policy Concept of the Russian Federation, approved by Presidential Decree No. 640 of November 30, 2016, states that the modern world faces the profound changes, the essence of which lies in the formation of a polycentric international system:

- "There is a disaggregation of the world's power and development potential, its shift to the Asia-Pacific region. The capabilities of the historical West to dominate the world economy and politics are being reduced . . . "(Art.4).
- "There is an aggravation in the contradictions associated with the world development disparity, the widening gap between the level of countries welfare, the intensification of the race for resources, access to markets, and control over transport corridors. The desire of Western states to maintain their positions, in particular by imposing their point of view on global processes and pursuing a policy of restraining alternative centers of power, leads to an increased instability of international relations, increased turbulence at the global and regional levels. The domination race in the key principles formation of the future organization of international system becomes the main trend of the current world development stage" (Art. 5).
- "In the context of aggravation of political, social, economic contradictions and the growing instability of the world political and economic system, the role of the factor of power in international relations increases. The buildup and modernization of the power potential, the development and implementation of new weapons' types undermine strategic stability, threaten the global security provided by the system of treaties and agreements in the field of arms control. Despite the fact that the danger of unleashing large-scale war, including nuclear war, among the leading states remains low, the risks of their involvement in regional conflicts and escalation of crises increase" (Art.6).
- "Existing military-political alliances are not able to provide resistance to the whole range of modern challenges and threats. In the context of the increased interdependence of all people and states, there are no longer any future attempts to ensure stability and security in a separate territory. The observance of the universal principle of equal and indivisible security for the Euro-Atlantic, Eurasian, Asian-Pacific and other regions becomes particularly important. There is a need in network diplomacy, which involves flexible forms of participation in multilateral structures for effective decision making of common tasks" (Art.7).
- "Important factors of state influence on international policy, such as economic, legal, technological and information are brought to the forefront, along with

military power. The wish to use appropriate options for the realization of geopolitical interests is detrimental to the search for ways of resolving disputes and existing international problems by peaceful means on the basis of the norms of international law" (Art.8).

- "The use of "soft power" tools, primarily the capabilities of civil society, information and communication, humanitarian and other methods and technologies, in addition to traditional diplomatic methods becomes an integral part of modern international politics" (Art.9).

In the *Foreign Policy Concept of the Russian Federation*, the position of the state in relation to the use of *information and communication technologies* (*ICT*) in the modern world is formulated. Thus, the document (Article 28) states that Russia takes the necessary measures:

- Ensuring national and international information security and counterthreats to state, economic, and public security originating from the information space
- Fighting against the terrorism and other criminal threats with the use of information and communication technologies (*ICT*)
- Counteracting the use of *ICT* for military and political purposes, including actions aimed at interfering in the internal affairs of states or posing a threat to international peace, security, and stability
- Seeking, under the auspices of the United Nations, for development of universal rules for responsible behavior of the states in the field of ensuring international information security, including the means of the internationalization of the information and telecommunications network of the Internet management on an equitable basis

At the same time, it is separately noted that Russia "conducts an individual and independent foreign policy course, which is dictated by its national interests and the basis of which is unconditional respect for international law. Russia is fully aware of its special responsibility for maintaining global security at both global and regional levels and is aimed to cooperate with all concerned states in the interests of solving common problems" (Art.21).

It is pointed out that Russia "places a high priority on ensuring the sustainable manageability of world development, which requires the collective leadership of the most powerful states. The leadership should be respectable to geographical and civilizational relations and be carried out with full respect for the central coordinating role of the United Nations" (Art.25).

It is noted that Russia "achieves its objective perception in the world, develops its own effective means of informational influence on public opinion abroad, helps to strengthen the positions of Russian and Russian-language media in the global information space, providing them with the necessary state support, actively participates in the international cooperation in the information sphere, takes the necessary measures to counter threats to its information security. To this end, it is expected that new information and communication technologies will be widely used. Russia will try to form a set of legal and ethical standards for the safe use of such technologies.

Russia defends the right of every person to access the objective information about events in the world, as well as the access to different points of view on these events" (Art.47).

1.1.2   Importance of the Information Space

Today, the international community recognized *land, sea, air, cosmos*, and the *information space* as integral components of the modern global world. It is important that the information space is equal to the components mentioned above. At the same time, the important remaining differences in the approaches of individual countries are the boundaries of the definition of the information space.

For instance, the USA uses a narrower term – *cyberspace* – which refers to some conditional (virtual) space that occurs during the use of computer facilities and data processing in computer systems and networks, as well as in related physical infrastructures [2–96]. The US Department of Defense defines cyberspace as a "global space in the digital environment, consisting of interdependent networks of information and communication infrastructures, including the Internet, communications networks, computer networks and embedded processors and controllers."[3] Defining the concept of *cyberspace* is the word cyber, which comes from the Greek κυβερνητκο'ζ and means the art of management. In Norbert Wiener's book *Cybernetics, or Control and Communication in Animal and Machine*, the term "*cybernetics*" was introduced in the context of control of complex systems. At present, this term has become widespread in all areas of human knowledge.

A group of experts from the Institute of West-East and the Institute of Information Security Problems (IPIB) of the Moscow State University named after M.V. Lomonosov proposed the definition of cyberspace as "*electronic (including photoelectronic, etc.) medium, in which information is created, transmitted, received, stored, processed and destroyed.*" According to Karl Rauscher, cyberspace does not exist without a physical component.[4] Here the cyberinfrastructure includes the following typical components:

- *Environment* (buildings, location of cell towers, satellite orbits, seabed where communication cables run, etc.)
- *Energy* (electricity, batteries, generators, etc.)
- *Hardware* (semiconductor chips, magnetic cards and printed circuit boards, wireline and fiber-optic data transmission systems)
- *Networks* (nodes, connections, network topology, etc.)
- *Transmission* (information transmitted via networks, statistics and traffic transfer schemes, data interception, data corruption, etc.)

---

[3] *Dictionary of military and related terms: US Department of Defense. – 2011. – P. 92–93*

[4] *Protection of communication infrastructure. Technical Journal of Bell Laboratories. – Special Issue: Internal Security. – Volume 9. – Issue 2. – 2004*

- *People* (engineers, developers, operators, maintenance personnel, etc.)
- *Policy* or in the expanded form of agreement, standards, policies and regulations

The Ministry of Defense of the Russian Federation defines the information space as "*a sphere of activity related to the creation, transformation and use of information, including individual and public consciousness, information and telecommunications infrastructure and information in particular.*"[5]

In the Information Security Doctrine of the Russian Federation 2016, the *information sphere* is understood as a total information, information facilities, information systems, sites in the information and telecommunications network of the Internet, communication networks, information technologies, entities whose activities are related to the formation and processing of information, the development and use of these technologies, information security, as well as a set of mechanisms for regulating relevant public relations.

1.1.3   Strategic National Priorities and Interests

The strategic goals and main directions of ensuring the information security of the state in the new Information Security Doctrine of the Russian Federation are determined to take into account the following Russian strategic national priorities:

- Ensuring the security of the country, its sovereignty, and territorial integrity, strengthening the rule of law and democratic institutions
- Creation of the external supportive environment for sustainable growth and improving the competitiveness of the Russian economy and its technological renewal, raising the level and quality of life of the population
- Strengthening the position of the Russian Federation as one of the most influential centers of the modern world
- Strengthening Russia's position in the system of world economic relations and preventing discrimination of Russian goods, services, and investments, using the capabilities of international and regional economic and financial organizations for this purpose
- Further course advancement toward strengthening international peace, ensuring universal security and stability in order to establish a just and democratic international system based on collective principles in solving international problems, on the supremacy of the international law, primarily on the provisions of the UN Charter, as well as on equal and partnership relations between states with the central coordinating role of the UN as the main organization regulating international relations

---

[5]*Glossary of terms and definitions in the field of information security:2nd ed., enlarged and revised. Military Academy of the General Staff of the Armed Forces of the Russian Federation. Research Center for Information Security. – M. – 2008. – P. 40*

- Enhancing the Russia's role in the global humanitarian space, spreading and strengthening the position of the Russian language in the world, popularizing the achievements of the national culture, historical heritage, and cultural identity of the peoples of Russia, Russian education, and science, and consolidating the Russian diaspora
- Strengthening the positions of Russian mass media and mass communications in the global information space and bringing the Russian point of view to international processes to the wider circles of the world community
- Promoting the development of constructive dialogue and partnership in the interests of enhancing the consensus and mutual enrichment of different cultures and civilizations, etc.

At the same time, national interests in the information sphere (Art.8 of the Doctrine) are:

A. Ensuring and protecting the constitutional rights and freedoms of a person and citizen in particular if they relate to the obtaining and use of information, the inviolability of privacy in the use of information technology, the provision of information support for democratic institutions, the mechanisms of interaction between the state and civil society, and the use of information technologies in the interests of preserving cultural, historical and spiritual, and moral values of the multinational people of the Russian Federation
B. Ensuring the stable and uninterrupted operation of the information infrastructure, primarily the critical information infrastructure of the Russian Federation and the unified telecommunication network of the Russian Federation, in peacetime, in the immediate threat of aggression, and in wartime
C. Development of the information technology and electronics industry in the Russian Federation, as well as the improvement of the activities of production and academic and scientific and technical organizations in the development, production, and maintenance of information security facilities and provision of information security services
D. Bringing to the international and Russian community the reliable information about the state policy of the Russian Federation and its official position on socially significant events in the country and the world, the use of information technologies to ensure the national security of the Russian Federation in the field of culture
E. Advancing the international information security system aimed at addressing risks to the use of information technologies by invaders for the purpose of violating strategic stability, strengthening an equitable strategic partnership in the field of information security, and protecting the sovereignty of the Russian Federation in the information space

It is significant that "the realization of national interests in the information sphere is aimed at the formation of a safe environment for the circulation of reliable information and the information infrastructure, which is stable to various types of influence, in order to ensure the constitutional rights and freedoms of a person and citizen, the country's stable social and economic development, as well as the national security of the Russian Federation" (Art.9).

1.1.4 Major Threats to Information Security

The approved Doctrine notes that "the opportunities of cross-border circulation of information are being increasingly used to achieve geopolitical, contrary to international law military-political, as well as terrorist, extremist, criminal and other unlawful goals, to the detriment of international security and strategic stability" [1, 6–20, 97–120].

The major threats to information security include the following:

- "The building up by a number of foreign countries the means of information and technical impact on the information infrastructure for military purposes. At the same time, the activities of organizations engaged in technical intelligence with respect to Russian state bodies, scientific organizations and military-industrial complex enterprises are intensifying" (Art.11).
- "The use of means for information and psychological impact aimed at destabilizing the domestic political and social situation in various regions of the world and leading to the undermining of sovereignty and violation of the other states territorial integrity by special services of individual countries. Religious, ethnic, human rights and other organizations are involved in such activity, as well as separate groups of citizens. The capabilities of information technologies are widely used for these purposes. There is a tendency to increase the volume of materials containing a prejudiced assessment of the state policy of the Russian Federation in the foreign mass media. Russian media is often exposed to outright discrimination abroad; Russian journalists are hampered by their professional activities. The informational impact on the, primarily youth, population of Russia is being increased in order to erode traditional Russian spiritual and moral values" (Art.12).
- "The use of mechanisms of information impact on individual, group and public consciousness by various terrorist and extremist organizations for the purpose of forcing ethnic and social tension, inciting ethnic and religious hatred or enmity, propaganda of extremist ideology, as well as attracting new supporters to terrorist activities. Such organizations actively develop means of destructive impact on the objects of critical information infrastructure for unlawful purposes" (Art.13).
- "The growth of computer crime, especially in the financial and credit sphere, increases the number of crimes related to the violation of constitutional rights and freedoms of a person and citizen, including those relating to privacy, person and family secrets, when processing personal data with a use of information technologies. At the same time, the methods, ways and means of committing such crimes are becoming more sophisticated" (Art.14).

- "The use of information technologies for military and political purposes, such as carrying out actions that are contrary to international law, aimed at undermining sovereignty, political and social stability, the territorial integrity of the Russian Federation and its allies and posing a threat to international peace, global and regional security" (Art.15).
- "The growth of cyber-attacks on critical information infrastructure facilities, the strengthening of intelligence activities of foreign states against the Russian Federation, and the growing threats of using information technology to aggravate the sovereignty, territorial integrity, political and social stability of the Russian Federation" (Art.16).
- "Insufficient level of competitive domestic information technologies development and their use for production and services. The level of dependence of the domestic industry on foreign information technologies remains high, in particular it relates to the electronic component base, software, computers and communications, which determines the dependence of the socio-economic development of the Russian Federation on the geopolitical interests of foreign countries" (Art.17).
- "Insufficient efficiency of scientific research aimed at developing prospective information technologies, low level of domestic developments implementation and insufficient human resourcing in the field of information security, as well as low awareness of citizens in matters of ensuring private information security. It is noted that measures to ensure the information infrastructure security, including its integrity, accessibility and sustainable functioning by domestic information technologies and domestic products often do not have an integrated basis" (Art.18).
- "The desire of individual states to use technological superiority for dominance in the information space. The current distribution among countries of the resources needed to ensure the safe and sustainable operation of the Internet does not allow the implementation of a joint, equitable, trust-based management. The absence of international legal norms regulating interstate relations in the information space, as well as mechanisms and procedures for their application that take into account the specifics of information technologies impedes the formulation of the international information security system aimed at achieving strategic stability and equal strategic partnership" (Art.19).

1.1.5 Strategic Goals and Main Directions of Information Security

In the *field of defense*, "the country protects the vital interests of the individual, society and the state from internal and external threats connected with the use of information technologies for military and political purposes that are contrary to international law, including for the purpose of carrying out hostile acts and acts of aggression aimed at undermining sovereignty, violating the territorial integrity of states, and posing a threat to international peace, security and strategic stability" (Art.20). Here, according to the military policy of the Russian Federation, the main IS directions are (Art.21):

A. Strategic deterrence and prevention of military conflicts that may arise as a result of the information technology use
B. Improvement of the Armed Forces of the Russian Federation information security system, other troops, military formations, and bodies, including forces and means of information confrontation
C. Forecasting, detection, and assessment of information threats, including threats to the Armed Forces of the Russian Federation in the IT field
D. Assistance in ensuring the protection of the interests of the Russian Federation's allies in the IT sphere
E. Neutralization of information and psychological impact, including those directed at undermining the historical foundations and patriotic traditions associated with the defense of the fatherland

In the field of *state and public security*, "the protection of sovereignty, the maintenance of political and social stability, the territorial integrity of the Russian Federation, the provision of basic human and civil rights and freedoms , as well as the protection of critical information infrastructure" (Art.22). Here the main directions of information security are (Art. 23):

A. Counteracting the use of information technologies for propaganda of extremist ideology, spreading xenophobia, ideas of national exclusiveness in order to undermine sovereignty, political and social stability, violent change of the constitutional system, and violation of the territorial integrity of the Russian Federation
B. Suppressing the activities that are detrimental to the national security of the Russian Federation, carried out with the use of technical means and information technology by special services and organizations of foreign states, as well as by individuals
C. Increasing the security of the critical information infrastructure and the stability of its operation, developing mechanisms for information threats detection and prevention, eliminating the consequences of their occurrence, and increasing the protection of citizens and territories from the consequences of emergency situations caused by information and technical influence on critical information infrastructure facilities
D. Improving the security of the operation of information infrastructure facilities, among others, to ensure a stable interaction of state bodies, to prevent foreign control over the operation of such facilities; to ensure the integrity, stability, and security of the unified telecommunications network of the Russian Federation; as well as to ensure the security of information transmitted through it and processed in information systems on the territory of the Russian Federation
E. Improving the security of the weapons, military and special equipment, and automated control systems functioning

F. Increasing the effectiveness of delinquency prevention, committed using information technologies and counteracting such violations

G. Ensuring the protection of information containing the data classified as state secret and other information with restricted access and dissemination, in particular by increasing the security of related information technologies

H. Improvement of ways and methods of production and secure application of products, service provision based on information technologies using domestic developments that meet the requirements of information security

I. Increasing the efficiency of information support for the implementation of the Russian Federation state policy

J. Neutralizing the information impact aimed at eroding traditional Russian spiritual and moral values

In the *economic sphere*, "the reduction to the lowest possible level of influence of negative factors caused by the insufficient level of development of the domestic industry of information technologies and electronic industry, the development and production of information security competitive means, as well as increasing the volume and quality of providing services in the field of information security" (Art.24). Here the main directions of information security are (Art.25):

A. Innovative development of the information technology and electronics industry, the increase of the share of this industry products in the gross domestic product, in the country's export structure

B. Liquidation of the dependence of the domestic industry on foreign information technologies and information security means through the creation, development, and widespread implementation of domestic developments, as well as the production and the provision of services on their basis

C. Increasing the competitiveness of Russian companies operating in the information technology and electronics industry, developing, manufacturing, and operating IS security facilities that provide services in this area, including by developing a supportive environment for operating in the Russian Federation territory

D. Development of the domestic competitive electronic component base and technologies for the production of electronic components, ensuring the demand of the domestic market for such products and the access of this product to the world market

In the field of *science and technology and education* – "supporting the innovative and accelerated development of the information security system, the information technology and electronics industry" (Art.26). Here the main directions of information security are (Art.27):

A. Achievement of competitiveness of Russian information technologies and development of scientific and technical potential in the IS field

B. Development and implementation of information technologies, initially resistant to different types of impact

C. Scientific research and pilot development in order to design perspective infor-
    mation technologies and means, providing information security
D. Development of human resources in the field of information security and
    application of information technologies
E. Security provision of citizens against information threats, including by creating a
    culture of private information security

In the field of *strategic stability and equal strategic partnership* – "the formation
of a stable system of non-conflict interstate relations in the information space"
(Art.28). The main directions of providing information security are (Art.29):

A. Protection of the Russian Federation sovereignty in the information space
    through the implementation of a separate and independent policy aimed at the
    realization of national interests in the information sphere
B. Participation in the formation of a system of international information security
    that provides effective counteraction to the use of information technologies for
    military and political purposes that are contrary to international law, as well as in
    terrorist, extremist, criminal, and other unlawful purposes
C. Development of international legal mechanisms that take into account the
    specifics of information technologies, in purpose of prevention and resolution
    of interstate conflicts in the information space
D. Promotion, within the framework of the activities of international organizations
    of the Russian Federation position, providing equal and mutually beneficial
    cooperation of all interested parties in the information sphere
E. Development of the national management system for the Russian segment of the
    Internet

The new Doctrine of Russian Federation Information Security was published
almost immediately after the introduction of the Concept of Russia's Foreign Policy.
In fact, the Doctrine is a development and continuation of the Concept with reference
to such a significant component of the country resources as the state information
space (including cyberspace). The main strategic goals of the new Doctrine: protection of Russia's sovereignty, maintenance of political and social stability of the
society, territorial integrity of the state, provision of basic human and civil rights and
freedoms, and protection of critical information infrastructure.

The preparation of the new Information Security Doctrine of the Russian
Federation was announced in March 2016. The previous Doctrine, adopted about
17 years ago, is visibly obsolete, as was noted by the Security Council of the Russian
Federation. It was necessary to develop a new document relevant to the current
developments in information and communication technologies (cloud, virtual,
hypervisors, IIoT/IoT, etc.). In addition, since the end of the previous century, the
list of immediate threats to information security has significantly expanded, relations
on the world scene have changed, and the possibilities of cyber-opposition have
become incomparably greater. Moreover, threats to information security reached the
level of interstate confrontation. At the same time, such concepts as information
operations, software-hardware, and psychological effects became a reality of modern

international relations. From the previous Doctrine, the document is characterized by brevity, clarity of presentation, and, at the same time, the breadth of the issues coverage under consideration. In particular, for the first time, there was formulated the goal of bringing to the Russian and international community an accurate information about the state policy of the Russian Federation and its official position on socially significant events in the country and the world, improving the protection of critical information infrastructure, neutralizing the impact of information aimed at the erosion of traditional Russian spiritual moral values (Art. 23), the elimination of the domestic industry dependent on foreign information technologies and IS means (Art. 25), supporting innovation, and ensuring the accelerated development of information security systems, the IT industry, and the electronics industry (Art.26).

According to Russian experts in the field of information security, this document will form the basis for the activities of public authorities, state, and most of commercial enterprises and organizations.

1.2   The Need to Monitor Cyberspace

The research's relevance in the field of intrusion detection is explained as follows: in a global network, thousands of cyber-attacks are carried out at all times and hundreds of fresh copies of malicious code (not detected by signature methods) are distributed, forming packages of infected workstations and servers (botnets). At their distribution peak, the most successful of them (e.g., Storm, Conficker) manage to control hundreds of thousands and, according to some estimates, even millions of computers. The vast majority can spread simultaneously through several different schemes, temporarily neutralize antivirus software, and bypass traditional computer detection systems by connecting to special control centers for instructions and "fresh" program code. What follows below are an analysis of salient security threats in 2017 and a discussion of the opportunities offered by modern methods of cyber-attack detection and prevention [121–124].

1.2.1   Security Threats Assessment

Malicious software developers primarily aspire to expand their own sphere of influence. Moreover, as in previous years, their efforts have tended toward one of two vectors – technical and social. It is worth mentioning upfront both of these trends were honestly "developed" and brought tangible benefits and considerable revenue to the hacker community.

1.2.2   Technical Direction

In 2017, the following main trends were observed as relates to exploiting vulnerabilities:

- The share of remote vulnerabilities in network software decreased, giving way to browser and office software vulnerabilities
- Within the category of browser vulnerabilities, there was a sharp increase in vulnerabilities in a variety of plug-ins and interface modules, despite a significant decline in the number of software errors in the browser core itself
- The rise in the exploitation of errors in productivity software, observed in 2016, continued in 2017
- The average number of newly detected remote and local vulnerabilities in the operating system code itself did not change significantly
- Greater attention was paid to the devices of network infrastructure and means of network protection as objects of vulnerability detection (Table 1.1)

Undoubtedly due to their widespread use, software products from Microsoft and Adobe remain by far the most common objects of attack, as concerns productivity software. Among Microsoft Office products, Office 2016 has the highest risks.

Table 1.1 Possible attacks on network infrastructure

Attack	%
Smurf	57.32215
Neptune	21.88491
Normal	19.85903
Satan	0.32443
Ipsweep	0.2548
Portsweep	0.21258
Nmap	0.04728
Back	0.04497
Warezclient	0.02082
Teardrop	0.01999
Pod	0.00539
guess_passwd	0.00108
buffer_overflow	0.00061
Land	0.00043
Warezmaster	0.00041
Imap	0.00024
Rootkit	0.0002
Loadmodule	0.00018
ftp_write	0.00016
Multhop	0.00014
Phf	0.00008
Peri	0.00006
Spy	0.00004

Although Office 2013 has a comparable number of vulnerabilities, its use has already sharply declined, thus curbing the potential spread of malicious code. As for the positive trends, it should be noted that Office 2010 products currently demonstrate an exceptionally low level of susceptibility to classic attack methods. That having been said, it remains unclear whether this is because potential intruders have not sufficiently investigated its code or the merit of the developers' having implemented tools and technologies to improve the software quality.

In contrast, unfortunately, Adobe software for viewing and editing digital documents and multimedia has preserved and even worsened its reputation concerning susceptibility to attack. For instance, multiple counts of very serious vulnerabilities have been identified, including some that permit the intruder to download and execute malicious code on the victim's computer. Furthermore, this heightened vulnerability was found across a broad range of products, both in terms of functionality and versions.

Competition "on the browser market" also has an increasingly noticeable impact on trends in intruders' search for vulnerabilities. Thus, the users of *Yandex*, *Google Chrome*, *Mozilla Firefox*, *Orbitum*, *Internet Explorer*, *Opera*, *Amigo*, and *K-Meleon* more often become the objects of successful vulnerabilities exploitation. Fortunately, the sharp increase in intruder interest in code errors in Mozilla Firefox has been offset by the developers' prompt response to the exposed vulnerabilities and the system of essentially "obligatory" auto-updates built into this browser. These factors jointly account for the very low level of real hacks. Nonetheless, as many analysts have noted, in 2017 set new records for the average time between a vulnerability publication of the vulnerability and its implementation in practice in malicious code spread throughout the Internet.

Malicious code developers are showing an increasing interest in network infrastructure objects: routers, DSL modems, DHCP servers, etc. For a lot of families of routers and modems running under one of the Linux clones (Fig. 1.1), a remote attack was developed, allowing to gain full control over the device and use it to further distribute the same code. For a known line of Cisco network firewalls, a vulnerability and a code were published that allowed an external attacker to transfer the device to an inoperable (before reboot) state. Still, malicious users are paying utmost attention to the SSH protocol and its specific implementations.

The technology of introducing fake DHCP servers into the local network was used more often and more successfully in 2017. The main purpose of the attackers was to substitute DNS servers and, as a result, to open wide opportunities for phishing and interception of confidential information. Moreover, hybrid technologies for spreading malicious code were improved, often combining completely different ways of penetrating a victim computer. An example of this approach is the modification of the WannaCry virus-worm so that after infecting the computer, it embeds malicious links into the HTML, ASP, and PHP pages found on the computer and network resources. When infected, pages were part of a www-server, which spread the threat not only to the owner of the pages but also to all his visitors.

The number of methods for introducing malicious codes at lower levels (new versions of rootkit technologies, implementation in the BIOS, microcode processors,

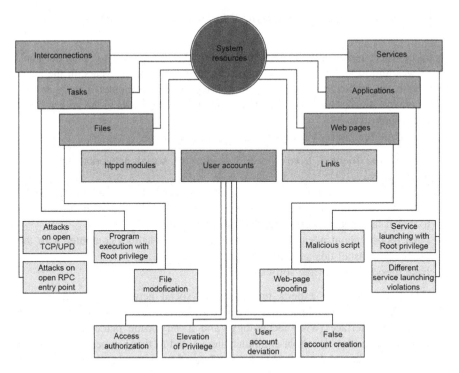

Fig. 1.1 Typical attacks on Linux clones

etc.) increased greatly. In this way, these developers hope to win control over the computer before downloading the antivirus software for passive or active counteraction to it (respectively – either hiding their code from the analyzer or neutralizing the antivirus subsystems, for example, the virus database update service). It should be noted that the increase in the prevalence of rootkits-trojans immediately affected the increased interest in antivirus boot disks (*LiveCD*), which are the most reliable in fighting with such malicious code.

The active resistance of Trojans to the antivirus products installed on the infected computer has improved significantly. Almost all "successfully" distributed modifications of virus programs are able to completely or partially disable antivirus software, block auto-update sites of virus databases, and counteract launched antivirus processes. Considering the confirmed facts of the active development of the market for automatic means of concealing malicious code (encryption, mutation, polymorphism) market, the malicious code of the observed processes can already be regarded as clashes in cyberspace between instances of "good" and "bad" code.

In general, the situation in the field of search and exploitation of new vulnerabilities and approaches to infect computers can be described as rather complicated. The openly published part of the datastream concerning new errors in the widespread software is not growing smaller, which could well be just the tip of the iceberg in comparison with the black market of vulnerabilities, accessible only to members of closed hacker groups.

1.2.3   The "Social Engineering" Direction

In cases where technical vulnerabilities did not help the attacker to reach the goal, there was almost always a way of social engineering. In recent years, at the same time with the predictable decline in the average level of computer literacy among new network users, it has developed into an industry comparable to the vulnerability detection industry.

The main directions of social engineering in 2017 were:

- Mailing to "friends" from hacked accounts of social networks and instant messengers
- Phishing e-mails with false notifications of new messages on social networks
- Fake interfaces of popular entertainment, news, and media sites with the proposal to install fake codecs or plug-ins for viewing
- Fake reports about alleged security threats on the victim's computer with the proposal to install fake antivirus products
- Fake messages about free software updates

Information security experts predicted that social networks and messengers turned out to be a "paradise" for social engineering methods. Subconsciously recognizable interface (www-technology) and a very wide percentage of users with medium and low levels of computer literacy led to rising cyberfraud in social networks. As it turned out, the percentage of transfers, even via clearly suspicious links and downloads that came through social networks from a well-known person, is several times higher than the figures typical for regular web surfing. The level of threats from social networks and messengers has grown to such an extent over the past year that even employers who are generally tolerant of their employees' web surfing began taking IS requirements seriously and completely blocking social networks and instant messengers on office computers.

Many leading experts have called the situation with fake codecs and plug-ins for web-viewing media content as one of the past year's biggest problems and an objective one at that. Modern software users have grown accustomed to receiving regular updates in order to correctly display all new media content specifications, and this provided intruders a perfect niche of fake codecs and plug-ins. Even an experienced PC user cannot always distinguish between a fake server source for the proposed update. At present, no decisive measures exist for reducing such risks.

Another related class of social engineering tricks is www-messages on virus programs which have supposedly been detected on the surfer's personal computer that flooded "gray" sites. Users are then given the recommendation to download a unique modern antivirus software, which triggers the installation of malicious code. The level of risk for this class of threats can be effectively reduced by a timely and high-quality educational program of "minimum knowledge on information security" for PC users. Unfortunately, this type of infection is rampant.

A new approach to phishing, brought in 2016 to workable incidents, was demonstrated by a symbiosis of malicious code developers and experts in websites promotion (*SEO*). The essence of it lies in the artificial rise of the rating for qualitatively made fake promo websites in the results on keyword search engines. It is not a secret that search engines are one of the main user guides to sites visited for the first time, and a high rating of a fake website gives an additional subconscious confidence in the website's safety. More oftenly, the result is an offer to install infected plug-ins or codecs, as described above.

The facts and schemes outlined above allow us to conclude that, recently, social engineering has become a powerful tool for developers of malicious code and is currently expanding rapidly.

1.2.4 What Is the Purpose?

What motivates attackers when they capture other PCs? Money and more money.

The key sources of income, and as a consequence, the objectives of infection, include:

- The use of infected computers for spam mailing
- Theft of the payment details stored on the computer's online payment systems (primarily – banking)
- Direct extortion of money (e.g., by encrypting files or blocking the computer operation)
- Theft of the details of social networks or instant messengers for phishing mailings on the contact list
- The use of infected computers for DDoS attacks in order to extort money

The year 2017 saw a redistribution in monetization schemes for criminal proceeds in connection with the appearance of a new "convenient for all" means of making payments via text messages. Certainly, the heightened cash flows via texts owe its inspiration to certain organizations designed to control the placement of paid and additional content in the cellular operator networks. Yet, the fact remains: the scheme proved viable and beneficial for intruders, effectively bringing them closer to their victims than ever before. In turn, this further tipped the scales toward the result, in which the owner of the infected PC will pay the extortionist via text message and in a few minutes will receive the unlock code, rather than seek the help of a malware removal specialist or a company system engineer (in the latter case, it might be because of fear of repressive measures from employer). It is no surprise that developers of file encryption, blockers of system operations, and authors of phishing mails actively switched to the new scheme.

1.2.5   What Does This Mean?

The general trends in the sphere of malicious code have largely mimicked those of previous years and are characterized by two key terms: "professionalization" and "monetization." The black market of malware contains offers of sale for infected PC arrays, descriptions of unpublished vulnerabilities, encryption programs for protection against antivirus signatures, programs for automatic generation of code "combines" that exploit the vulnerability and malicious content. Developers of antivirus software and systems of detection and prevention of intrusions and anomalies reduce the response time to new virus instances to the greatest extent possible. However, the signature approach is fundamentally reactive, and proactive (including heuristic) methods have still not achieved the desired correlation of detection level and percentage of false responses. At the same time, the high bandwidth of communication channels and the widely spread distribution algorithms in global networks allow malicious code to overtake huge computer parks in a matter of hours and days.

In this light, we can say that the situation with cyber-attacks remains rather complicated, and unfortunately the year 2017 did not offer any tangible methods in the fight against it. For this reason, research in the field of detection and countering of computer attacks is of high priority and has drawn close attention from specialists.

We will discuss below the main ways to detect and counter intrusions into computer networks and systems.

1.2.6   The Ultimate Capabilities of Known Methods to Fight Cyber-attacks

At present, the issues of axiomatics, terminology, methodology, and connection of the theory of detecting and connecting cyber-attacks along with other scientific disciplines of proper information security remain in a formative stage.

Of the existing works on the taxonomy of detection methods for the abovementioned type of attacks, two works were particularly emphasized that appeared almost simultaneously [21, 89]. In the first, in spite of the detailed classification of other accompanying characteristics (such as information sources, detection response, etc.), the actual solving algorithms are divided only in two broad classes: with a prior knowledge of the system (*knowledge-based*) and behavioral (*behavior-based*) (Fig. 1.2). In the second work, the classification already takes into account both the possibility of self-learning and training with the teacher in behavioral systems and also classifies signature classes with an independent space allocation for the key attributes.

In the work of M. Almgren [18], the authors attempted to bring up the compliance tables for terms and classes of the key published works and proposed classification of approaches in the form of a two-dimensional plane in the axes "self-learning-prior knowledge" and "deviation control-norm control."

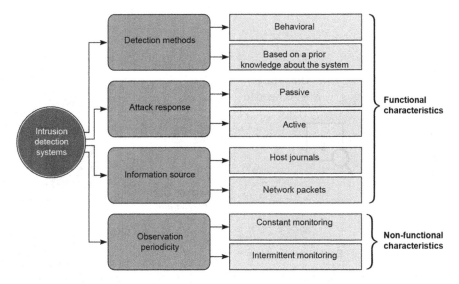

Fig. 1.2 IDS classification according to H. Debar

A detailed summary of the classification issues of antivirus systems and application systems for intrusion detection, partially addressing the classification of the algorithmic and methodological base, is presented in the work of *IBM Research Zurich* [17, 80] and *RTO/NATO* [18, 57, 125] (Figs. 1.3 and 1.4).

1.2.7   Traditional Methods Review

Two classification systems have a significant influence on the properties of the method groups for cyber-attack detection: level of the data processed and the decision-making scheme, concerning the violation existence (decision scheme algorithm).

Classification by the level of the processed data divides the methods into the analyzers:

L.1) Binary representation of data or command codes
L.2) Commands, operations, events, and/or their parameters (without regard to their physical representation in computer facilities)
L.3) System characteristics, directly or indirectly reflecting its intended purpose, for example, statistics of involved resources, the number of requests processed per unit of time, speed and other characteristics of network exchange, etc. (Table 1.2)

Algorithms of low-level (machine-dependent) analysis are usually much simpler in implementation, have a higher speed, and are the least demanding of all resources. Examples of algorithms of class L.1 are [17, 18]. On the other hand, the analysis of

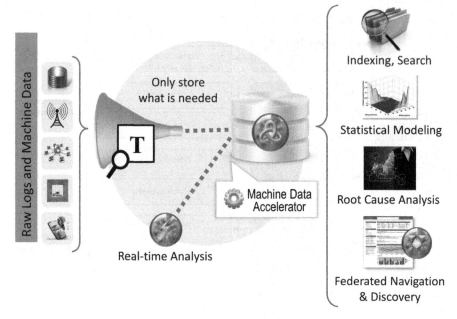

Fig. 1.3 Data process levels in IDS, IBM Research Zurich

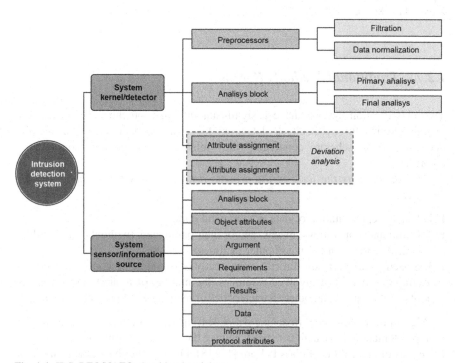

Fig. 1.4 IDS, RTO/NATO algorithm breakdown

Table 1.2 Levels of the processed data

Levels of the operational environment	Label	Bypass methods	
Level 7. Tasks (work)	TASK	Program execution masking	Local
Level 6. Programs	JOB	Difficulties, connected with the program analysis on the application layer	
Level 5. Program processes	PS	Use of system library spoofing	
		Interception of system calls	
Level 4. System calls and interrupts	SVC	Change of import process table	
		Spoofing of the export table	
Level 3. Command system	ISP	Spoofing of interrupt processor	
Level 2. Interconnection processes "processor–memory"	PMS	Introduction of changes in command computer code	
Level 1. Schemes and register transfers	CRT		
Levels of OSI model	**TCP/IP protocols**	**IDS bypass methods**	
Level 7. Application layer	HTTP, SMTP, SNMP, FTP, Telnet, SCP, NFS, RTSP	Custom coding	Network
		Traffic encoding	
		Polymorphism	
Level 6. Presentation layer	XML, XDR, ASN.1, SMB, AFP	Difficulties, connected with the program analysis on the application layer	
Level 5. Session layer	TLS, SSH, ISO 8327/CCITT X.225, RPC, NetBIOS, ASP	TTL value manipulation	
		IP fragmentation	
		TCP fragmentation	
Level 4. Transport layer	TCP, UDP, RTP, SCTP, SPX, ATP	TCP sequence order manipulation	
Level 3. Network layer	IP, ICMP, IGMP, X.25, CLNP, ARP, RSRP, OSFP, RIP, IPX, DDP, BGP		
Level 2. Channel layer	Ethernet, Token ring, PPP, HDLC, Frame relay, ISDN, ATM, MPLS	Broadcasting mailing	
		Multiple VLAN headers	
		Insignificant frame oversize	
Level 1. Physical layer	Physical environment and information encoding principles		

the two higher levels provides decision algorithms with a more targeted flow of information, which potentially improves the quality of the decisions made at the same computational power cost while also providing a certain degree of platform independence. Algorithms of class L.2, for example, include [126].

The highest level of analysis of the system state (class L.3) [32, 126] usually gives indirect information concerning existing deviations, which often necessitates the involvement of an operator in order to discover the true cause of the abnormal system operation. However, in some cases this is the only information source about the ongoing malicious impact (e.g., in a distributed denial of service attack by generating a large number of correct but resource-intensive queries or in similar cases).

Classification according to the decision-making scheme on the presence of abnormal system operation seems to be the most suitable from the positions of the pattern recognition theory, to which the given problem generally belongs (Table 1.3).

D.1. *Structured methods* of recognition form a rigorous model of either a knowingly correct state or impact or deliberately malicious influence. Other variants of impacts, including possibly correct or malicious (but unknown at the time of model creation), are not analyzed and lead to errors of type I or II, depending on the chosen analysis policy. Using these class methods offers the advantage that there are no false responses in the area governed by the model; yet, the disadvantages are the fundamental impossibility of describing new or unknown previously methods of malicious influences as well as those that do not fit into the developed model.

D.1.1. *State correctness control.*

D.1.1.1. *Inspection algorithms* perform the most rigorous system control. They check the file integrity [34, 126, 127](implemented in Tripwire – AIDE – and similar systems), memory areas, or more complex data structures (e.g., network routes prefix bases) based on records of their knowingly correct state: the file size, checksums, cryptographic hash sums, and so on.

D.1.1.2. *The algorithms of the system state graph/transition graph control or protocol model* represent the most widely discussed subclass of structured methods. The analysis is carried out on significant events occurring in the system, whose current state is known. By describing the authorized transitions for each state, it becomes possible to generate events when the system behavior deviates from the authorized one. One of the first works in this direction was study [128] implemented in *STAT* and *USTAT* systems, respectively. Subsequently, a subclass of methods was identified, which uses *Petri nets* to control the sequence of events [129]. At present, research is being conducted on the possibility of increasing the description flexibility of permissible system behavior [123, 130–132] and of automating the process of constructing the authorized transition graph.

D.1.1.3. *The monitoring algorithms of the standard exposure policy* are a complete or partial description of the authorized impacts on the system, thereby forming a "default deny" policy, presuming that any attempt to violate it constitutes an informative event. Along with the inspection algorithms, there is a subclass that has the greatest history in the field of cyber-attack detection. Various options for

Table 1.3 IDS-related literature reference list

Year	Article	Description	Class	Level	Point
2006–2017	usenix 2006lad.pdf	Informing real owners about fake BGP routes	Structural/inspector	Command	Network
2003–2017	usenix 2003\pennington.pdf	Anomaly search system (comparison with calculated before checksums by data storage monitoring)	Structural/inspector	Byte	System
2003–2012	immune_overview.pdf	Immune system review (no specifics)	Correlation/immune	Command	Network
2002–2006	carleton\2002_balthrop.pdf2002_balthrop_2.pdf	Immune system, uses 49-bit hash as a detector including receiver IP, sender IP, and TCP port number	Correlation/immune	Command	Network
1999–2005	an_artificial_immune_model.pdf	Detailed immune system analysis as COA	Correlation/immune	Command	Network
2002–2017	liao02use.pdf	Anomaly search in system call sequence by clustering	Correlation/cluster	Command	System
2005–2016	vigna\2005_valeur_mutz_vigna_ dimva05.pdf	Anomaly probability calculation by Markov chains in the moment of www-application queries to database (SNL query analysis) to receive the information	Correlation/cluster	Command	No data
2004–2012	DOECSG2004.pdf	Formation of anomalies that are the closest to boundary detection clusters on the basis of genetic algorithms (extensive learning is required)	Correlation/genetic cluster	Command	Network
2005–2017	gam\54.pdf	Neural networks by number of packets, arrival intervals, etc.	Correlation/neural	Command	Network
2004–2010	gam\76_mukkamala04\intrusion.pdf	Decision-making based on neural networks and SVM	Correlation/neural	Command	Network
2005–2017	http://www.iisi.msu.ru	Vasyutin S.V., Zavyalov S.S./neural networks: analysis of call statistics for a certain period of time	Correlation/neural	Statistics	Network

(continued)

Table 1.3 (continued)

Year	Article	Description	Class	Level	Point
2005–2017	http://www.iisi.msu.ru	Reich V.V., Sinitsa I.N., Sharashkin S. M./neural network ART2M	Correlation/neural	Statistics	System
2007–2017	carleton\compnet_2007.pdf	Construction of a finite automaton of frequently encountered tokens in www-queries	Correlation/behavioral	Command	Network
2002–2008	soma_diss.pdf	Dissertation on pH system (COA for Linux based on the profile analysis of system function call sequence)	Correlation/behavioral	Command	System
2002–2016	carleton\tan_why6.pdf	Anomaly search system description based on the statistics analysis of 6-letter sequences, given as system call parameters	Correlation/behavioral	Byte	System
1995–2001	gam\04_5sri.pdf	NIDES statistics application (statistics of the system calls and involved resources)	Correlation/threshold	Statistics	System
1997–2006	gam\55_jou00design.pdf	JiNao is a statistic profile, signatures, and state machine for OSPF	Correlation/threshold	Statistics	Network
2006–2016	carleton\kruegel.web.pdf	Anomaly search in www-queries by multiple parameters extraction from URL, decrease of dimensions, and then weighing on the graph	Attribute/cluster	Command	Network
2004–2015	carleton\stolfo_payl.pdf	Anomaly search in www-queries based on the Mahalanobis distance evaluation for the occurrence rate of specific symbols in URL-queries	Attribute/cluster	Byte	Network
2005–2012	gam\87_nvat_ua_aiccsa.pdf	Multidimensional statistical analysis of the attribute space	Attribute/cluster	Statistics	Network
2005–2017	http://www.iisi.msu.ru	Petrovsky M.I/application of methods of intellectual	Attribute/cluster	Command	Network
		Data analysis in computer intrusion detection problems (FuzzySVM)			

2006–2017	raid2006AAAI07.pdf	Worm search in the subsystems by global subsystem analysis to find more than four incoming connections per 50 s with the following clustering and rejection of false responses	Attribute/threshold	Statistics	Network
2005–2012	dn\containment.pdf	Ever-fast detection and isolation of the worms in response to scanning of the surrounding machines	Attribute/threshold	Statistics	Network
2004–2016	dn\portscan_oakland04.pdf	Decisive rule of scanning detection based on mathematical theory	Attribute/threshold	Statistics	Network
1988–2000,2016	gam\94_smaha.pdf	Statistical behavior profile	Attribute/threshold	Statistics	Network
2007–2017	http://www.spbstu.ru	Baranov P.A./cluster compactness evaluation	Attribute/threshold (on the cluster volume)	Statistics	System
2007–2017	carleton\ndss07_xssprevent.pdf	Analysis of different XSS fight methods by analysis of Java code or execution process in Java machine	Structured/transition graph	Command	System
1998–2004	gam\48x_frincke98planning.pdf	Transition graph analysis, however with the partial set of state and transition conditions	Structured/transition graph	Command	System
1995–2006	gam\52_ilgun95state.pdf	Transition graph analysis	Structured/transition graph	Command	System
1997–2011	gam\55_jou00design.pdf	JiNao is a statistic profile, signatures, and state machine for OSPF	Structured/transition graph	Command	Network
2007–2017	vigna\2007_balzarotti_cova_felmetsger_vigna_multimodule.pdf	Review of the different methods of www-application scripts statistical analysis (data flow, control, links, etc.) to search for module deviation	Structured/graph model	Command	System
2006–2016	usenix 2006kirda.pdf	Detection of spyware, embedded in browser helper objects through MinAPI calls (policy deviation)	Structured/policy	Command	System

(continued)

Table 1.3 (continued)

Year	Article	Description	Class	Level	Point
1996–2004,2017	gam\57_specification_based.pdf	Description of the profile of the system command call sequence in the stage of program installation and its further check	Structured/policy	Command	System
2004–2017	gam\90_sheynerthesis.pdf	Determined finite automation	Structured/policy	Command	System
1988–2012	gam\94_smaha.pdf	User policy	Structured/policy	Command	System
2004–2017	http://www.cs.msu.su	Gamayunov D.Yu./finite automation description for the process	Structured/policy	Command	System
2004–2008	usenix 2004\kim.pdf	Automatic signature generation system for newly detected worms	Structured/signature	Byte	System + network
2006–2012	raid2006\anagram_raid2006.pdf	Automatic signature generation system for newly detected worms in tapped data	Structured/signature	Byte	Network
2006–2014	carleton\kruegel_polyworm.pdf	Unique code print formation for poly-morphic worms on the basis of control transfer graph analysis	Structured/signature	Byte	System + network
2004–2017	carleton\singh_earlybird.pdf	Automatic signature generation for new worms	Structured/signature	Byte	Network
1997–2006	gam\55_jou00design.pdf	JiNao is a statistic profile, signatures, and state machine for OSPF	Structured/signature	Command	Network
1994–2004, 2016	gam\60_kumar94pattern.pdf	Detection of incorrect action sequences of partial compliance with malicious impact signature	Structured/sig-nature (unclear)	Command	System

implementing this approach were applied in a variety of access control systems (including one of Haystack's first fully functional IDS in 1988).

D.1.2. Monitoring (search) of nonstandard impacts.

D.1.2.1. Algorithms for monitoring nonstandard policies constitute a description of the list of deliberately prohibited influences on the system, forming a "default allow" policy. Unlike monitoring of the standard impact policy, which can be derived from a protocol or some formal description of the desired system behavior, forming a complete list of prohibited impacts is difficult, and in many cases impossible, because the information systems are complex and multi-level. The decisive class rules are free from "false triggering" error, which gives them a significant advantage in implementing without the operator participation in systems. However, they are not able to detect new types of malicious influences on the system that are not accounted for in their knowledge base, and consequently, the quality of their work largely depends on the speed of updating the attacker model.

D.1.2.2. Signature algorithms search for previously known patterns of computer intrusions and differ in the level of analysis (according to the classification given above) and in the degree of template specification/generalization. Algorithms in this class use antivirus software products, as well as network traffic filtering systems (including mail and web content). Modern research within this class at the level of the analysis of commands/events are primarily focused on the universalization of knowledge bases aimed at unifying information updates relating to attacks of different etymologies, levels, and intensity and also at the scalability of related systems. In a subclass of methods performing search at the byte-oriented level, research is being conducted in the field of automatic generation of intrusion signatures, as well as in the search for effective methods of countering mimicry and polymorphism in the attacks (for example, by analyzing the transition graph in the worm binary code [123, 131–133]).

D.2. Correlation methods introduce metrics to distinguish the observable attribute vector or more complex (for example, behavioral) characteristics from a state known to be either correct or malicious. They are characterized by the fact that they form certain (positive or negative) values for the entire set of impacts. It is also applied to extremely unlikely conditions (even the reliability degree of decision-making in them is insignificant). The advantage of correlation methods is the coverage of the entire set of permissible impacts, which hypothetically allows us to make correct decisions with respect to previously unknown attacks. For this class of algorithms, it remains, however, difficult to cumulatively reduce the level of first and second type errors. With respect to all correlation methods, both implementations in the "learning with the teacher" mode and in the self-learning (adaptive mode) are possible.

D.2.1. Algorithms "without memory" consider each event (impact, system transition from one state to another, or one measurement indicator of any system characteristic) as a separate element of the set, in relation to which it is necessary to make a decision. The term "attribute space methods" is also applicable to this class.

D.2.1.1. *With one-dimensional attribute vector.*

R.2.1.1.1. *Threshold algorithms* generate information events on the fact of anomaly detection when the observed value exceeds a certain boundary value. Threshold algorithms were the first representatives of the class of intrusion detection correlation methods; in particular, they are described in paper [6], which was fundamental for the entire area under consideration in Haystack in 1987. The most widely used in practice was the control of the system volume of the requested resources and the control of the certain event frequencies in the system (e.g., for a specific type of events, in research works [123, 126, 131–133], aggregated for dispersion statistics of the frequency vector – in the study [126]).

D.2.1.2. *With multidimensional attribute vector.*

D.2.1.2.1. At present, *the algorithms of linear classification* in multidimensional attribute space have ceded the way to the more flexible algorithms of cluster and neural network analysis.

D.2.1.2.2. *Cluster analysis* as a proven method of classification received wide application in the field of cyber-attack detection. At present, research is exploring the possibility of detection without a teacher (search for significant deviations) and clustering with preliminary training on marked input data [123, 131, 132].

P.2.1.2.3. *Neural network methods* are used to make decisions on the presence or absence of malicious influence of a decisive scheme based on the neural network. The first works in this class are dated back to the late 1990s [32, 126, 131, 132, 134]. Currently, the number of different methods in the class is quite large, including independent domestic research. In particular, in [126] it is proposed to apply neural networks of adaptive resonance, and in work [117, 134] the decision is made by the neural network on the basis of the vector, containing the frequencies of the system queries and the state identifier of the controlled computing process.

D.2.1.2.4. *Immune methods* [123, 131, 132] attempt to extend the principles of detection and counteraction of the immune system of living beings to foreign viruses. The system includes a centralized "gene library" that form a limited set of vectors, characterizing potentially alien events and a distributed system of sensors that perform actual detection of the effects and have a feedback coupled to the "gene library." The methods are characterized by the undemanding nature of resources, but in some circumstances, they generate a high flow of false events.

D.2.2. *Algorithms "with memory"* analyze events taking into account some history and, also probably, the true or assumed state of the system.

D.2.2.1. *Deterministic behavior control algorithms* generate events for any fact of the system's deviation in behavior from the profile created at the training stage and are somewhat analogous to the inspecting algorithms in the class of structural methods. In case of an unsuccessful selection of the protection object or a list of monitored events, they can generate a high false positive rate. Fuzzy algorithms are generally preferred for their flexibility.

D.2.2.2. *Fuzzy behavior control algorithms*, in the course of analyzing the sequence of events in one way or another, compute a vector of probability characteristics and generate an event only when they exceed certain threshold values. The analysis is possible both at the byte level (e.g., in [29, 126] the parameters of the system queries are analyzed) and at the command/event level [13].

These main research trends have the common aim of achieving the following results:

- Improving the accuracy of the decision-making algorithms (reducing the levels of type I/II errors, especially with respect to previously unobserved effects, both malicious and correct)
- Increasing the proportion of corrective processes (Figs. 1.5 and 1.6) that do not require the involvement of a human operator, thus bringing response time to

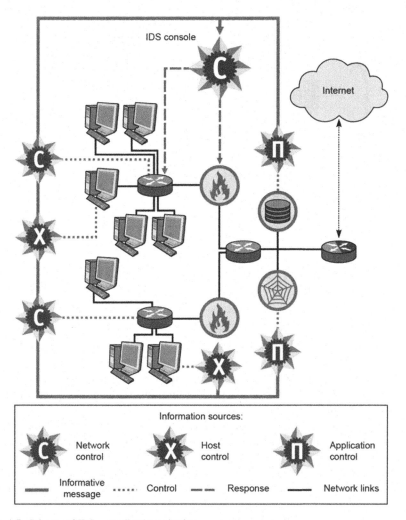

Fig. 1.5 Scheme of IDS centralized monitoring

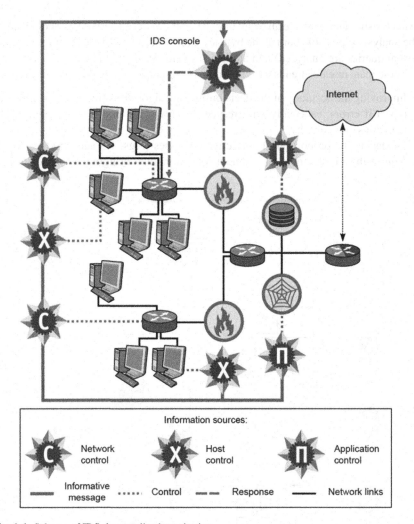

Fig. 1.6 Scheme of IDS decentralized monitoring

malicious impact to an all-new level (e.g., automatic generation of signatures for new malicious code a few minutes after confirming its abnormally rapid propagation through the network)
- Counteracting the new technologies used by intruders in order to hide the fact of harmful effects, for example, using polymorphic encoders of executable code and data or mimicry techniques ("dissolution" or masking in normal traffic) of attacks or generate an active impact on antivirus protection by creating conditions for denial of service or generating an excessive flow of false responses, rendering its application impossible (Figs. 1.7 and 1.8).
- Generate an active impact on antivirus protection by creating conditions for denial of service or generating an excessive flow of false responses, rendering its application impossible

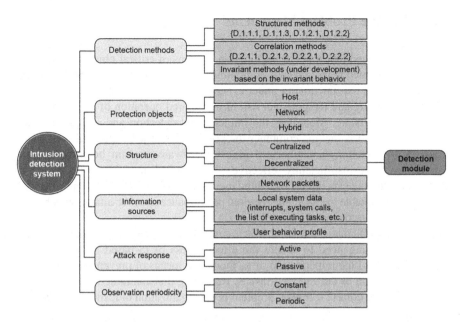

Fig. 1.7 IDS possible classification

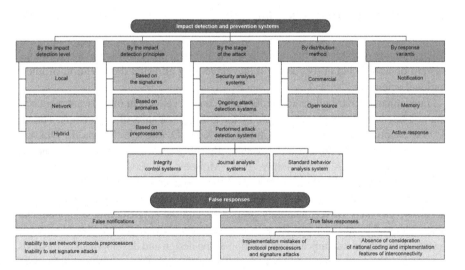

Fig. 1.8 Taxonomy of the response systems to computer attacks: (**a**) impact detection systems; (**b**) classification of false responses of the impact detection systems

1.3   Possible Problem Statements

It is acknowledged that modern information systems of critical objects (which are complex distributed heterogeneous computing systems) do not possess the required stability for targeted functioning under cyber-attack conditions because of the high design complexity and the potential risk of undeclared operation of equipment and system-wide software. Moreover, the implemented detection and complex cyber-attack neutralization method combines the possibilities of joint application of technologies for obtaining unauthorized access, hardware-program bookmarks, and malware remain insufficiently effective. The applied methods and resilience methods, using the capabilities of redundancy, standardization, and reconfiguration, are no longer suitable to ensure the required performance in cyber-attacks which lead to disastrous consequences.

This is a problematic situation, which content lies in the contradiction between the ever-increasing need to ensure the effectiveness of IS-based systems during cyber-attacks and the imperfection of methods and means of prevention, detection, and neutralization of destructive impacts. Resolving this contradiction requires the resolution of the current technical problem is the organization of resilient computing in IS systems under cyber-attack [127, 135, 136].

1.3.1   State-of-the-Art Review

The problems of ensuring the reliability and stability of the software and technical system operation and co-occurring issues have long attracted the attention of international scientists.

Fundamental contributions to the formation and development of software engineering as a scientific discipline were made by the eminent scientists of our time: A. Turing, J. Von Neumann, M. Minsky, A. Church, S. Klini, D. Scott, Z. Manna, E. Dijkstra, Hoar, J. Bacus, N. Wirth, D. Knuth, R. Floyd, N. Khomsky, V. Tursky, A.N. Kolmogorov, A.A. Lyapunov, N.N. Moiseev, AP Ershov, V.M. Glushkov, A.I. Maltsev, and A.A. Markov. They laid the foundation of theoretical and system programming, enabling the analysis of possible computing structures with mathematical accuracy, the study of computability properties, and the simulation of computational abstractions of feasible actions. Russian scientific schools were founded based on these results. They made a significant contribution to the design and development of methods to increase the program reliability and resilience through the automatic synthesis of model abstractions to specific software solutions.

The Siberian Branch of the Russian Academy of Sciences made an important contribution to the development of the program schemes theory (A.P. Ershov, Y.I. Yanov, V.E. Kotov, V.K. Sabelfeld), the formation of analytical and applied verification, methods of correctness proof and transformational synthesis programs (V.A. Nepomnyashchii, O.M. Ryakin, D.Ya. Levin, L.V. Chernobrod), the study of fundamental properties of algorithms and universal programming (B.A. Trakhtenbrot, V.N. Kasyanov, V. A. Evstigneev), and the study of abstract data types and denotational semantics (V.N. Agafonov, A.V. Zamulin, Yu.L. Er Seam, Yu.V. Sazonov, A.A. Voronkov). It opened the principle of mixed computations and laid the foundation of concrete programming (A.P. Ershov).

Institute of Cybernetics of Ukraine named after V.M. Glushkova initiated algebraic programming (Yu.V. Kapitonova, A.A. Letichevsky, E.L. Yushchenko), compositional programming (V.N. Red'ko), structural schematics, macro-conveyor calculations and automata-grammatical synthesis of programs (V.M. Glushkov, EL Yushchenko, GE Tseitlin), production P-technology and multi-level design, and the survivability of computing systems (A.G. Dodonov). At the same time, to prove the completeness, correctness, and equivalence of programs, the apparatus of V.M. Glushkov systems of algorithmic algebras (SAA) was proposed.

The Institute of Cybernetics of Estonia brought to life the ideas of automatic program synthesis in a workable Programming System with Automated Program Synthesis for Engineering Problem solution (PRIZ system) and developed the field of conceptual programming and NUT technologies (E.H. Tyugu, M.Ya. Harf, G.E. Minz, M.I. Kakhro). The Latvian State University achieved significant results in the field of inductive synthesis of programs, symbolic testing, and methods and tools of program verification and debugging (Ya.M. Barzdin, Ya. Ya. Bichevsky, Yu.V. Borzov, A. A. Kalninsh).

The Moscow and St. Petersburg Academic Schools gave to the modern software technology the fundamental ideas for the automatic synthesis of knowledge-based programs (D.A. Pospelov), intellectual knowledge banks and logical application computing (L.N. Kuzin, V.E. Wolfengagen), intellectual programming (V. Strizhevsky, N. Ilyinsky), synthesis of programmable automata (V.A. Gorbatov), generators of programs – GENPAK (D. Ilyin) and intellectual problem-solvers (Yu.Ya. Lyubarsky, E.I. Efimov), programming in associative networks (G.S. Tseiti), reliable software design (V.V. Lipaev), automated testing of software based on formal specifications (A.K. Petrenko, V.P. Ivannikov), hyper programming (E.A. Zhogolev), synthesis of abstract programs (S.S. Lavrov), the synthesis of recursive metamodels of programs (R.V. Freivald), metaprogramming in problem environments (V.V. Ivanishchev), active methods for increasing the reliability of programs (M.B. Ignatiev, V.V. Filchakov, A.A. Shtrik, L.G.Osovetskiy), the approach to the design of absolutely reliable systems (A.M. Polovko), the sign modeling methodology (Yu.G. Rostovtsev), and many more.

Scientists of the Military Space Academy named after A.F. Mozhaysky contributed significantly to algorithmic design and development (R.M. Yusupov,

V.I. Sidorov), information (Yu.G. Rostovtsev, B.A. Reznikov, S.P. Prisyazhnyuk, A.K. Dmitriev), technical (A.M. Polovko, A. Ya. Maslov, V.A. Smagin), and program reliability (Y.I. Ryzhikov, A.G. Lomako, V.V. Kovalev, V.I. Mironov, R.M. Yusupov).

However, it is necessary to pose the problem of ensuring the performance of an IS system under cyber-attack in a new way, so that the organization of the restoring computation while under mass attack would prevent significant or catastrophic consequences. In a way similar to a living organism's immune system, is the proposed solution intends to give the computer system the ability to develop immunity to disturbances in the computational processes while under the conditions of miscellaneous massive attack. In order for this to become possible, a solution must be found to the new, relevant, urgent scientific problem of computational auto-recovery in a state of mass disturbance.

The main goal here is to provide the required level of stability for critical IS systems computing undergoing cyber-attacks. To achieve the goal, it was necessary to put and solve the following scientific and applied problems:

- Technical analysis of the problem of maintaining the operational resilience of IS systems undergoing mass cyber-attacks
- Investigation of ways to ensure the computational resilience of these systems during group and mass cyber-attacks
- Development of the concept of ensuring computational stability for group and mass disturbances
- System representation of organizing resilience to disturbances computations
- Modeling of typical disturbances and development of scenarios for their neutralization
- Formalization of correct calculation semantics
- Development of detecting disturbance methods and restoring partially correct computations
- Development of a technique for detecting and neutralizing cyber-attacks on IS systems

The possible scientific and methodological research apparatus consists of the following:

- Methods of dynamic systems for modeling computations in the conditions of destructive actions of intruders
- Methods of the multilevel hierarchical systems' theory for the system design of a resilient computing organization
- Methods of the formal languages theory and grammars for generating possible types of scenarios for dissimilar mass disturbances and for recognizing the structures of matched effects

- Methods of theoretical and system programming for modeling computational structures and computational operations for group and mass disturbances aimed to synthesize an abstract program for auto-recovery of disturbed calculations
- Catastrophe theory methods for analyzing the dynamics of the disturbed calculations behavior by analogy with modeling disturbances in living nature
- Similarity theory methods for proving computability properties of reconstructed computations
- Methods of the monitoring and computation recovery theory for developing immunity to destructive impacts

The scientific novelty of the results, conclusions, and recommendations is that the following were proposed and developed for the first time:

- Concept of ensuring the stability of the functioning of IS systems with the system of immunities based on auto-recovery computation capable of auto-recovery
- Scientific and methodological apparatus to organize resilient computing under destructive cyber-attacks, including:

  - Model of computation organization resistant to destabilization in a hierarchical multi-level monitoring environment with feedback coupling
  - Models of the simplest disturbances for the scenarios synthesis of the return of computational processes to equilibrium using dynamic equations of catastrophe theory
  - Model of representing correct computational semantics based on static and dynamic similarity invariants
  - Methods for auto-recovery of computing with memory using permissive standards
  - Technique for detection and neutralization cyber-attacks of the "denial of service" type, DDOS applying the system of immunities

The research's practical significance largely owes to the fact that using the immunity system makes it possible to develop and accumulate measures for counteracting previously unknown cyber-attacks, detecting group and mass impacts that lead to borderline-catastrophic states, partially recovering computational processes which provide the solution to target system problems based on IS systems which prevent their degradation, and wielding unrecoverable or difficult to recover disturbances against the attacker.

1.3.2 Problem Formalization

The research task statement in the functions is as follows (Inset 1.1).

Inset 1.1 Mathematical Statement of the Problem of Resilient Calculation Organization

Given:

Calculation organization system in IS under disturbances

$\Theta = (T, U, Y, X, F_p, F_v, E$, where

T is a set of the observance time moments of computational process

U, Y – state space of resilient calculations with the recovery under the disturbances

X – state space of computational process under mass disturbances

F_p – organization operator of resilient calculations with the recovery under the disturbances

F_v – immunity memory formation operator

E – accumulated protection immunity against the mass disturbances

Functions to be found:

1. $l \div T \times U \times X \rightarrow Y$
2. $\eta \div Y \rightarrow E$
3. $v \div Y \rightarrow E^*$
4. $\chi \div E \rightarrow E^*$
5. $\psi \div E \rightarrow Y^*$
6. $\zeta \div E \rightarrow Y^*$

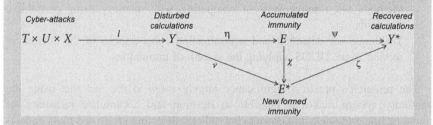

l – calculation disturbance function

η – function of calculation state compliance of accumulated immunity

v – function of calculation state compliance of new formed immunity

χ – function of immunity base updating

Ψ – function of calculation recovery based on accumulated immunity

ζ – function of calculation recovery based on new formed immunity

To determine the work constraints, the following stability indicators were used: the probability of the output information presentation P_i in a given period of IS operation T_g, the probability of performing technological operations P_t in a given period of IS operation T_g, the average time of the technological operation executionT_e or the probability of performing the technological operation P_e in a given time T_g, the probability P_c of the correct result obtaining the information processed in a given time T_g, and the probability of neutralization of the dangerous programmatic impact of P_n in a given period of IS operation T_g.

Thus, it was suggested to maintain the operability of IS systems when the target tasks are solved under intruder cyber-attacks by recovering the computation processes based on the accumulated "immunities" not after the attack but while the disturbances are underway.

1.3.3 Possible Solutions

A model of an abstract computer with memory with the use of the formal apparatus of R.E. Kalman theory of dynamical systems is proposed. In general form, the model of an abstract computer under disturbances $\Re$ with a discrete time, m inputs and p outputs over the field of K integers, is represented by a complex object ($\aleph$, $\wp$, $\Diamond$), where functions $\aleph : 1 \rightarrow 1$, $\wp : K^m \rightarrow l$, $: l \rightarrow K^p$ are core abstract K-homeomorphisms, and l is some abstract vector space over K. Space dimension $l(dim\ l)$ determines the dimension of the system $\Re$ (dim $\Re$). The chosen representation made it possible to formulate and prove the assertions confirming the fundamental existence of the desired solution in the paper.

In accordance with this, the ideology of computation with memory was introduced and disclosed for imparting immunity to disturbances. The general presentation of the immune system of the IS providing stability of its behavior under mass and group intruder impacts is considered on the following representation strata (Fig. 1.9).

Firstly, monitoring of cyber-attacks and accumulating immunity are performed, consisting of the following:

- Modeling cyber-attacks in types of destructive impacts
- Modeling the representation of the computational disturbances dynamics and determining scenarios for returning the calculations to the equilibrium (stable) state
- Developing a macro model (program) of auto-recovery of computations under mass and group disturbances

Secondly, the program development and verification for auto-recovery of disturbed calculations on the microlevel are aimed at the following:

- Developing a micro model (program) of auto-recovery of computations under mass and group disturbances
- Modeling by means of denotational, axiomatic, and operational semantics of computations to prove the partial correctness of recovered computations' computability features

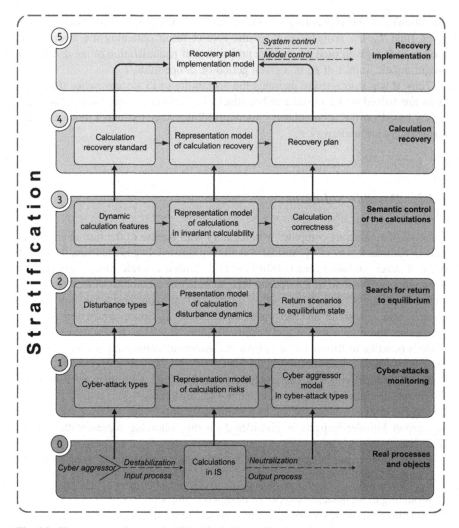

Fig. 1.9 The representation strata of IS critical object self-recovery processes

And, thirdly, auto-recovery of the disturbed calculations is achieved in the solution of target tasks at the microlevel:

- Development of operational standards for computation recovery
- Development of a model for their representation
- Development and execution of a plan for computation recovery

Thus, the abstract solution of the formulated scientific problem is presented in a general form in the types of theoretical models that allow synthesizing the required computational structure with memory, to prove the partial correctness of the computability features and to design an abstract auto-recovery plan.

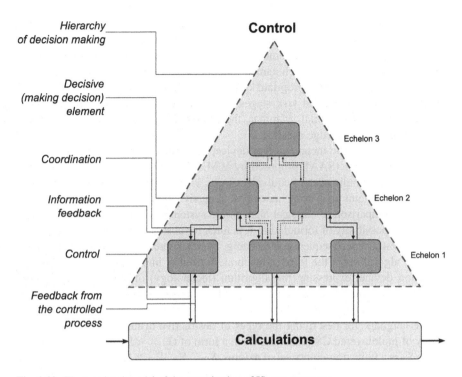

Fig. 1.10 The two-level model of the organization of IS auto-recovery

To synthesize programs for the auto-recovery of disturbed computations, it was necessary to develop a system of knowledge that allowed us to describe the technology of generating auto-recovery tasks in accordance with the stages of setting the problem, planning its solution, and its subsequent implementation. In accordance with this, three information models and one model for managing the process of solving problems were formally defined. In this case, the information model in the problem statement made it possible to describe potential types of disturbances in computations. The decision planning model made it possible to reflect the cause-effect relationships between the objects of the abstract auto-recovery program. The solution implementation model made it possible to describe the internal structure of the auto-recovery program. Such two-level modeling (Fig. 1.10) enabled a degree of foresight regarding the solution to the global auto-recovery problem of disturbed computations through a sequential solution of local subtasks.

$$(\forall y)(\forall x)\{[P(x,\bar{D}(y))\text{и}\,Q_0(y,x)] \Rightarrow [P(x,\bar{D}(y))\text{и}\,P(\pi_M(x),D)]\}$$
$$(\forall y)(\forall x)\{[P(x,\bar{D}(y))\text{и}\,Q_0(y,x)] \Rightarrow P(\pi_M(x),D)\}$$

The proposed multimodal approach differs fundamentally from the known single-model approaches and permits description of abstract auto-recovery programs for disturbed calculations in the structural-functional, logical-semantic, and pragmatic aspects. Such a multi-model system for organizing the management of computation auto-recovery required the implementation of coordination, thus permitting consideration of the specificity of each named functional model, which, in turn, led to the necessity of designing an appropriate knowledge metamodel. Based on the structure and content of the selected task stages as formalisms of the basic models, it is advisable to use the formal grammar, the production system, and the automatic converter in the knowledge system.

While choosing the metamodeling device, preference is given to the system of algorithmic algebras (SAA), proposed by V.M. Glushkov, which enabled the creation of an algorithmic system equivalent in its visual capabilities to such classical algorithmic systems as Turing machines, recursive functions, and Markov algorithms. In comparison to classical algorithmic systems, this approach's advantages include the possibility of formulating structures for abstract auto-recovery programs in the Dijkstra types (sequence, branching, cycle), representing the required auto-recovery algorithms in the form of algebraic formulas, improving the formal transformation device, expressing the algorithm (technology) of auto-recovery in elementary operators, and effectively transforming auto-recovery programs for disturbed computations in machine implementation.

The language for describing the types of destructive disturbances is given by a family of multilayered CS grammars G_p in a form of $G_p = <S_G, N_G, T_G, R_G, P_G>$, where S_G is a finite non-empty set of axioms, $S_G \subseteq N_G, |S_G| \geq 1; N_G = \{a_i \mid i \in I_N\}$ is a finite non-empty set of types of information operations of the cyber aggressor (nonterminal), $T_G = \{x_i \mid i \in I_T\}$ is a finite non-empty set of types of cyber-attacks (terminals), $N_G \cap T_G = \varnothing; R_G = \{r_i : a_i \rightarrow \beta \mid i \in I_R, \alpha_i, \beta_i \in (N_G \cap T_G)^*\}$ is a finite set of inference rules, where $X_G = \{x_i \mid i \in I_x\}$ is a finite set of attributes, $(N_G \cap T_G) \rightarrow X_G$ is a bijection, and $P_G = \{p_i(.) \mid i \in I_p\}$ is a finite set of predicates, X_G. $G_p : g : M_g \times X^{g,j} \rightarrow G_p, g = <g_T, g_N, g_{R^0}, g_{R^P}, g_{X^0}, g_{X^P}, g_P, g_S>$. On this basis, a method has been developed for generating combinations of joint types of cyber-attacks, leading to miscellaneous mass destructive IS disturbances. To recognize the structure types of these destructive impacts, a method is proposed that allows one to make conclusions based on the data on the facts registration of group and mass IS disturbances with the help of recognizer family with stock memory.

Thus, a two-level model of IS auto-recovery organization that allows formal description of the structures and content of technological and procedural models of knowledge representation is necessary and sufficient for the generation of scenarios of various disturbance types of computations and is proposed and justified for the synthesis of abstract programs for their partially correct self-recovery.

Further, possible models of typical destructive impacts and scenarios of returning the IS state to equilibrium are based on catastrophe theory. The possibility analysis of natural model application of mass disturbances for the organization of computing with memory is carried out. The simplest natural models of mass disturbances and their canonical forms are considered (see Table 1.4).

Table 1.4 The canonical form of the equations of typical natural disturbances

K	n	Canonical form $f(x, a)$	Title
1	1	$x_1^3 + ax_1$	Folding
2	1	$x_1^4 + a_1\dfrac{x_1^2}{2} + a_2 x_1$	Gather operation
3	1	$\dfrac{x_1^5}{5} + a_1\dfrac{x_1^2}{2} + a_2\dfrac{x_1^2}{2} + a_3 x_1$	Swallow tail
4	1	$\dfrac{x_1^6}{6} + a_4\dfrac{x_1^4}{4} + a_1\dfrac{x_1^3}{3} + a_2\dfrac{x_1^2}{2} a_3 x$	Butterfly
3	2	$x_1^3 + x_2^3 + a_3 x_1 x_2 - a_1 x_1 - a_2 x_2$	Hyperbolic umbilical point
3	2	$x_1^3 - 3x_1 x_2^2 + a_3\left(x_1^2 + x_2^2\right) - a_1 x_1 - a_2 x_2$	Elliptical umbilical point
4	2	$x_1^2 x_2 + x_2^4 + a_3 x_1^2 + a_4 x_2^2 - a_1 x_1 - a_2 x_2$	Parabolic umbilical point
5	1	$x_1^7 + a_1 x_1^5 + a_2 x_1^4 + a_3 x_1^3 + a_4 x_1^2 + a_5 x_1$	Teepee
5	2	$x_1^2 x_2 - x_2^5 + a_1 x_2^3 + a_2 x_2^2 + a_3 x_1^2 + a_4 x_2 + a_5 x_1$	Second elliptical umbilical point
5	2	$x_1^2 x_2 + x_2^5 + a_1 x_2^3 + a_2 x_2^2 + a_3 x_1^2 + a_4 x_2 + a_5 x_1$	Second hyperbolic umbilical point
5	2	$1 \pm \left(x_1^3 + x_2^4 + a_1 x_1 x_2^2 + a_2 x_2^2 + a_3 x_1 x_2 + a_4 x_2 + a_5 x_1\right)$	Symbolic umbilical point

The analysis of regularities of the behavior of natural systems under the disturbances by their phase portraits is carried out. The conditions for the existence of possible trajectories of the disturbed natural systems' return to equilibrium are determined (see Fig. 1.11).

The application of natural models with the canonical form of disturbance dynamic representation of the IS self-recovery is substantiated. The similarity between destructive impacts of intruders and natural models of the catastrophe theory is revealed and formally described, which made it possible to define a meta-program for controlling the auto-recovery of disturbed IS states. The conditions for returning the disturbed IS states to an equilibrium state were determined. Analyzing the possibility of using natural models of mass disturbances for the organization of IS auto-recovery made it possible to determine the requirements for the meta-control via the process of auto-recovery (Fig. 1.12).

It is meant that for k control parameters $a_1, a_2, \ldots, a_k$ calculation parameters take such values $x_1^*, x_2^*, \ldots, x_n^*$ in a state of equilibrium in which local function minimization is achieved $f(x_1, x_2, \ldots, x_n; a_1, a_2, \ldots, a_k)$. In the general case, the values of x_1^*, corresponding to the equilibrium state, depend on the choice of parameters α so $x_1^*(\alpha), i = 1, 2, \ldots n$.

A representation of the disturbed computations is proposed in the following form. For each $k \leq 5$ and $n \geq 1$, there exists an open dense set C^∞ of potential functions F such that C_f is a differentiable k-variety that is smoothly imbedded in R^{n+k}. It is shown that each feature of the catastrophe mapping $s: C_f \to R^k$ is locally equivalent to one of a finite number of standard types, called simple disturbances or elementary catastrophes. The function m is structurally stable at each point M_f in relation to small f from F.

Fig. 1.11 Example of the canonical return to equilibrium scenario

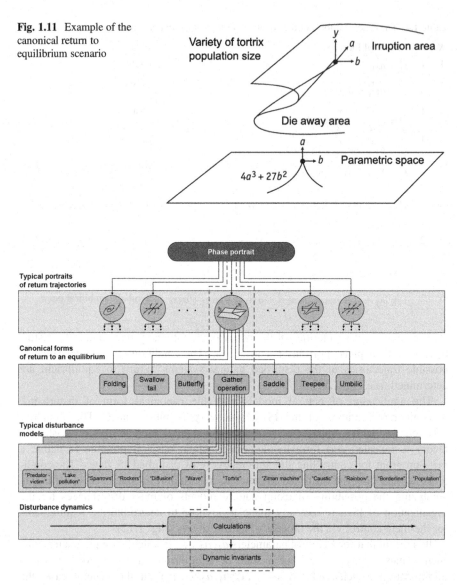

Fig. 1.12 Technology of detecting similarity between destructive impacts on IS and natural models of catastrophe theory

To determine the scenario of returning disturbed computations, the canonical forms of return to equilibrium were investigated. In particular, we propose the presentation of disturbed computations by equations of a saddle point of the form (Inset 1.2).

Inset 1.2 Destructive Impact

$y^3 + ay + b = o$, where

$$a = \frac{\left[\gamma^2 \alpha_1 \alpha_2 \bar{E}^2 \left(\alpha_3 + \bar{E}^2\right) + \gamma \alpha_2 \alpha_4 \bar{E}^3 - \gamma^2 \alpha_1^2 \alpha_2^2 \bar{E}^6\right]}{a\left(\alpha_3 + \bar{E}^2\right)}$$

$$b = \frac{\left\{\dfrac{-\frac{2}{27}\gamma^3 \alpha_1 \alpha_2^3 \bar{E}^9}{\left(\alpha_3 + \bar{E}^2\right)^2} + \gamma \alpha_3 \bar{E}^3 \dfrac{\gamma^2 \alpha_2 \alpha_5 \bar{E}^2 \left(\alpha_3 + \bar{E}^2\right)}{3\left(\alpha_3 + \bar{E}^2\right)} - \gamma^3 \alpha_1 \alpha_2 \alpha_5 \bar{E}^5\right\}}{a\left(\alpha_3 + \bar{E}^2\right)}$$

Here the critical branches in space a-b, where discontinuities may occur, satisfy the eq. $4a^3 + 27b^2 = 0$. Necessary and sufficient conditions for the existence of returning trajectories are determined. Scenarios for returning disturbed computations to equilibrium are synthesized.

The semantics of return-to-equilibrium scenarios for computations with memory are specified. The structure of the representation of the calculation return scripts with memory is defined. The similarity conditions between return scenarios and procedures for disturbed calculation recovery are established. The program structure for auto-recovery of disturbed computations by control operators of an abstract computer is proposed.

Thus, the similarity conditions between return scenarios and procedures for recovery of IS disturbed states were established. The structure of programs for self-recovery of disturbed IS states by controlling operators of an abstract computer is described. The methods of memory formation with the trajectories of the IS disturbed states return based on the results of solving a system of dynamic equations in the canonical form are considered.

Then, a language for semantics representation of partially correct IS states in terms of similarity theory was developed. Semantics representations of partially correct computations in similarity invariants were presented. The syntax of invariant dependencies was described for expressing schemes of computability invariants in functional forms. The similarity invariants were interpreted for the identification and elimination of semantic contradictions.

The conditions for the partial correctness of auto-recovery computations were determined. The system of relations between dimensions and invariants of computation similarity was analyzed to reveal the features of the formation of computability invariants. We propose the creation of computability invariants by abelian groups of similarity invariants to determine the partial correctness of the control structure of the self-recovery program. The invariant features of computability by graphs were investigated to determine the completeness and consistency of the operation structure of the self-recovery program. The criteria of the semantic correctness of computations to control key properties of computability were determined.

When choosing the apparatus for the structure modeling and computation features, a constructive combination of methods for analyzing dimensions and similarity invariants was proposed. The relation systems between the dimensions of input and output computation parameters $[\varphi_{us}(x_1, x_2, \ldots, x_n)] = [\varphi_{uq}(x_1, x_2, \ldots, x_n)]$ (here the entry [X] means "dimensionality of the value of X"), which allow each operator containing computational operations and/or an assignment operator (and having a homogeneous form with respect to its arguments) to represent each function as a sum of functionals φ:

$$f_u(x_1, x_2, \ldots, x_n) = 0, \quad u = 1, 2, \ldots, r,$$

$$\text{where } f_u(x_1, x_2, \ldots, x_n) = \sum_{s=1}^{q} \varphi_{us}(x_1, x_2, \ldots, x_n) = \prod_{j=1}^{n} x_j^{a_{jus}} \text{ were explored.} \quad (1.1)$$

As a result, it became possible to form the requirements for the dimensions of the quantities x_j and synthesize the similarity invariants of the computations as follows:

$$[\varphi_{us}((x_1, x_2, \ldots, x_n)] = [\varphi_{uq}((x_1, x_2, \ldots, x_n)], \quad (1.2)$$

$$\left[\prod_{j=1}^{n} x_j^{a_{jus}}\right] = \left[\prod_{j=1}^{n} x_j^{a_{juq}}\right]$$

$$\prod_{j=1}^{n} [x_j]^{a_{jus}} = \prod_{j=1}^{n} [x_j]^{a_{juq}}$$

$$\prod_{j=1}^{n} [x_j]^{a_{jus} - a_{juq}} = 1$$

After the reduction to a linear form, for example, by logarithm, we obtain a system of equations

$$\sum_{j=1}^{n} (a_{jus} - a_{juq}) \cdot In[x_j] = 0, u = 1, 2, \ldots, r; s = 1, 2, \ldots, (q-1). \quad (1.3)$$

In this case, the required criterion for the semantic correctness of the computations is the existence of a system solution in which none of the variables $((\ln[x_j]))$ is zero.

The representation of the similarity invariants is generalized as a finite set G with elements g_i, $i = 1, 2, \ldots, n$. On the G set, a binary operation $\oplus$ is defined which by a certain rule assigns to each pair of elements g_1, $g_2 \epsilon G$ an element $h \epsilon G$, denoted by $h = g_1 \oplus g_2$. It is shown that a certain operation $\oplus$ has closure properties (if $g_1, g_2 \epsilon G$, then $h = g_1 \oplus g_2$ is an element of the G set) and associativity (if $g_1, g_2, g_3 \epsilon G, h_1 = g_2 \oplus g_3, h_2 = g_1 \oplus g_2, h_1, h_2 \epsilon G$, then $h_1 \oplus g_1 = h_2 \oplus g_2$) and permits finding the unit element (G contains the left (right) unit e such that for each element $g \epsilon G$, $g \oplus e = e \oplus g = g$) and the inverse element ($g \epsilon G$ in G there is an inverse element g^{-1} such that $g \oplus g^{-1} = q^1 \oplus g = e$). This makes it possible to represent a set of invariants for the similarity of the computations of G with a binary operation $\oplus$ (algebraic addition) by a finite abelian group. It is defined that for the structural transformations of the similarity invariants of computations, the transformation superposition plays the role of the addition operator, and for the functioning condition transformation of the computational process, the composition of the transformations plays the role of the multiplication operator. It is shown that the groups W (the group of structural transformations of the similarity invariants) and V (the group of transformations of the conditions for the functioning of the IS) can be "embedded" in G, thereby defining subgroups G_W and G_V:

$$G_V = \{g = (e, w), w \in W, g \in G\}, \{g = (h, e), h \in V, g \in G\} \qquad (1.4)$$

where W is the group of structural transformations of the similarity invariants and V is the group of transformations of the conditions for the IS functioning.

Thus, the obtained results make it possible to study the computation semantics under disturbances and present it in the form of an appropriate system of dimensional equations and similarity invariants of computations. A procedure for synthesizing similarity invariants of computations is presented as follows: the choice of the algorithm implementation trajectory of the computational process; the verification of the sample representability over the covering of all the control graph vertices; the reduction of the control graph to separate the computational operators from the condition checking and loop organization operators, extracting from the sample all unique operators satisfying the restrictions described above on the form of the functional connection; the choice of the variables and constants of the operators under consideration; and the selection and presentation of similarity invariants for semantically correct computational processes. It is assumed that the elements of data array lie in the same dimension, and the numerical constants are values of pairs from different dimensions (the determination of their belonging to certain classes will occur automatically at the stage of the dimension matrix reconciliation). The control graph transitions, associated with the calculation of complex functional dependencies, or correspond to operators of subprogram statement, calls and complete the system with sets of operations of value assignment to formal parameters. In this way, it became possible to construct a new relationship between the calculation parameters. In this way, it became possible to construct a new relationship between the calculation parameters.

To verify the auto-recovery program of the disturbed computations, Floyd's inductive method was used, which made it possible to verify the truth of the output predicate in the truth chain of all intermediate implications. In this case, three groups of variables were defined: the input vector $u = (u_1, u_2, \ldots, u_n)$ consisting of the abstract auto-recovery program's input variables, software vector $x = (x_1, x_2, \ldots, x_n)$ applied to indicate the temporary memory used during the self-recovery program, and output vector $y = (y_1, y_2, \ldots, y_n)$ specifying output variables at the end of the self-recovery program. Accordingly, three (non-empty) domains are considered: the input D_u, the program D_x, and the output D_y. We consider the input predicate H $(u) : D_u \to \{I, L\}$ that defines elements of D_u that can be used as input variables of the auto-recovery program and the output predicate $j(u, y) : D_u \times D_y \to \{I, L\}$ describing the relations that must be fulfilled between the variables of the auto-recovery program after its completion.

As a result, it was proved in the paper that the program of auto-recovery of disturbed P computations has been completed, if for $\forall u$, such that $H(u)$ is true, the execution of the given program is completed by reaching the final state. The program for auto-recovery of disturbed P computations is partially correct with respect to H and j, if for $\forall u$, such that $H(u)$ is true and the program terminates, $j(u, y)$ is also true, and is partially correct (correct) according to H and j, if for $\forall u$, such that $H(u)$ is true, the execution of the program is completed and $j(u, y)$ is true.

To annotate the scheme of the auto-recovery program for the disturbed P computations, the input predicate H, and the output predicate j, it was proposed to successively apply the following steps: cutting the cycles, determining the appropriate set of inductive assertions, and determining the verification conditions. It thus follows that if all the verification conditions are true, then the auto-recovery program for the disturbed P computations is partially correct with respect to H and j. The proof of the partial correctness of the auto-recovery program is reduced to verification of each path in terms of given predicates and proving the verification truth of the conditions themselves.

To prove the completeness of the auto-recovery program, assertions were made about the variable state of the abstract program at some of its points, the verification conditions for the specified program were generated and brought to the annotated form, the consistency of the verification conditions were proved, and the auto-recovery program were completed. In particular, a finite number of cutting points for the auto-recovery program cycles were chosen, each of which is associated with some a priori true assertion $q_i(u, y)$. Then the assertion was tested

$$\forall u \left[H(u) \bigwedge R_\alpha(u) \supset g_i(u, r_\alpha(u)) \right] \tag{1.5}$$

for each α path from the initial point to the i point and

$$\forall u \forall x \left[g_i(u, x) \bigwedge R_\alpha(x, u) \supset g_i(u, r_\alpha(u, x)) \right] \tag{1.6}$$

for each α path from point i to point j, where r_α is the transformation performed on the α path; R_α is the assertion formulated on this way. Next, a choice of a well-ordered set $(W, \backslash prec)$ is made. There is a certain associated with each set point partial function $f_i(u, x)$, which takes $D_u \times D_x \to W$ and

$$\forall u \forall x \left[g_i(u, x) \supset f(x, u) \in W. \right] \tag{1.7}$$

Finally, the feasibility of the auto-recovery program termination for each α path from point i to point j was tested.

To automate the generation stages of condition verification according to the annotated program and to demonstrate the truth of the obtained formulas, Z, Mann's method was used in combination with Hoare axiomatics that enabled the introduction of desired assertions according to various types of auto-recovery programs for disturbed computations.

The condition truth evidence for the control structure correctness of the auto-recovery program is given. Derived rules of inference in calculus of computability invariants are proposed to form the conditions for the correctness of the logical structure and properties of the auto-recovery program. The invariants of computability to matrix representations are introduced by means of equivalent transformations. The consistency and completeness of the calculation of the computability invariants are established to achieve the desired expressiveness of the calculus. The independence and solvability of the calculus of computability invariants are defined to ensure uniquely complete genera.

A method of the feature proof of partial correctness and completeness of auto-recovery computations is proposed. For this purpose, the predicate calculus is modified, which allowed obtaining the assertion proof in the combined logic of the invariants of computability and similarity. The abstract execution of annotated programs in the notation of logical calculus is simulated. An annotated self-recovery program was synthesized with identically true pre- and postconditions on the operators of its control and executive structures. A verbal execution of an annotated auto-recovery program on an abstract computer was performed to prove the completeness of the selected structure.

Thus, the semantics of auto-recovery computations are formalized, thus making it possible to prove the partial correctness and potential completeness of the recovered IS states.

In conclusion, the final design scheme of an abstract program for the auto-recovery of disturbed IS states was presented. To this end, we have developed feasible operator designs that make up the body of the control operators of the auto-recovery program. There are many types of permissive standards for computations with memory that specify the recovery procedure. We establish the equivalence classes of the computability invariants of auto-recovering computations and their correlation with a system of solvable standards. The content of permissive standards for calculations with memory is based on the falling return path to equilibrium. The technology of designing internal bodies of the control structures of the self-recovery program was developed.

For this purpose, a model system developed earlier for the synthesis of abstract programs for auto-recovery of disturbed computations based on SAA was used. The structured model proposed (grammatic, product system, and automaton) allowed obtaining basic automata constructs implementing the process of synthesis of auto-recovery programs for disturbed computations.

The structured model application required different forms of data representation. Therefore, there was a need to find a general mechanism for the expression of declarative, declarative-procedural, and procedural data and their joint use. As such a mechanism, it was suggested to use the algebra of data structures (*ADS*) based on *SAA*, oriented on recognition and generation of initial, resulting, and intermediate data. The obtained algebra of data structures belongs to the number of polybasic algebraic systems and is a pair of base sets $<O^*, P^*>$ with a certain operation signature Δ, where $O^* = O/O \subset F(T)$ is a set of objects, $T_A = S \cup W$, $S \cap W = \varnothing$, S is a set of processed data, W is a set of delimiters, and $F(T_A)$ is a set of all finite sequences of symbols (configurations) in the alphabet T_A; $P^* = = \{a/0, 1, \mu\}$ is a set of three-valued logical conditions. In the signature Δ, the appropriately interpreted operations of the SAA signature are included, as well as some special operations on data structures. For ASD<O^*, P^*>, there is a $\prod$basis, consisting of elementary objects $O = \{0_i, i = 1, \ldots, n\}$ and elementary logical conditions $P = \{a_i, i = 1, \ldots, m\}$.Here, an abstract data type (ADT) is an algebraic system $\langle M_N, \Delta \rangle$, where M_N is a support represented by the corresponding OPC and Δ is a signature of the operations and predicates defined on M_N.

By introducing the ADT concept, it became possible to express the data structures and rules for their processing in SAA schemes, respectively, as OPC and RS. The construct of the data representation is memory in an abstract, logical, and physical sense, which has an access mechanism with a fixed set of operations. Here the abstract memory type (AMP) is a system, $A^* = \langle N_{mu}, S_a \rangle$, where N_{mu} is a support (the set of memory units); S_a is an access signature (multiple OPC in some $\prod$ basis, consisting of a finite set of elementary operators and conditions). A natural generalization of the memory structure with dense data placement is the elastic tape (ET), with which many widely distributed memory structures can be represented, such as a sequential file, a store, list structures, variable length strings, and many others.

The AMT (n) was proposed: an automaton with input and output channels, as well as channels for working with n internal tapes, which is defined by the object:

$$A = \langle S, X_A, Y_A, Z_i, \varphi_A \varphi_i \psi_i, \delta_l, s_0, F \rangle, \tag{1.8}$$

where S is a finite set of automaton states, $s_0 \in S$ is the initial state, $F_s \subseteq S$ is a set of final states, X_A and Y_A are the input and output alphabets, respectively, Z_i is an alphabet of the i-internal tape, and $\varphi_A : S\{X_A \times \Lambda\} \rightarrow S$is a transition function associated with the input chain reading (the Λ symbol is used to switch the machine without accessing the input tape); $\varphi : S \times Z_i \rightarrow S$ is a transition function associated with reading from the cell of the i-internal tape; $\Psi_i : S \rightarrow S \times Z_i$ is a transition function associated with writing to the τ-cell of the i-inner tape; $S \rightarrow S \times Y_\Lambda$ is an output function.

At each instant of discrete automatic time, the deterministic AMT (n) automaton performs one of the five types of elementary actions: reading from the input tape, reading and writing to the i-cell of the internal tape, writing to the output tape, and changing the state without accessing the tapes.

To represent AMT (n) automata oriented to the formalization of structured processes over standardized memory in PC terms, it is suggested to apply a control grammar (C-grammar) that is regarded as the object

$$G = \langle X_B, Y_B, Z, R \rangle$$

where X_B, Y_B are input and output alphabets, $Z = \bigcup_{i=1}^{k} Z_1$is a combined alphabet of internal tapes $L = \{L_i/i = 1, \ldots, k\}$, and $R = \{r_i/i = 1, \ldots, n\}$ is a set of product complexes, among which the axiom C-grammar r_1 and the final complex r_n. The complex of C-products has the form:

$$m : \alpha F \left(\coprod \right) \beta \prod \left(\frac{B_{ij}}{r_{ij}} \right), \tag{1.10}$$

where m is a label of the C-product; α is an applicability condition; β is a condition of performance correctness; $F(\coprod)$- PC, functioning over n AMT; and $\prod \left(\frac{B_{ij}}{r_{ij}} \right)$ is a switcher that transmits control by the B_{ij}conditions on complexes r_{ij} of receiver production. The meaning of the C-product consists in the implementation of the PC $F(\coprod)$ over a basis (the access signature to the corresponding AMT) under the truth of the applicability condition with the subsequent transition to complexes of C-products. In this case, the derivation in the C-grammar begins with the application of the complex r_1 and ends with a transition to the final complex r_n. For the C-grammar, the assertion is proved that an equivalent C-grammar G can be designed for each AMT (n)-automaton, and for each C-grammar G, an equivalent AMT (n)-automaton can be synthesized.

In the class of C-grammars arbitrary recursively enumerable sets could be represented that determines the fundamental possibility of design of an algorithmic language using the C-grammar, which provides an unambiguous description of the AMT automaton operation. The development of such a language allowed the joint description and subsequent synthesis of the declarative, logical, and procedural components of the abstract self-recovery program.

Thus, basic constructs were found for the expression of typical models of automata for setting the disturbance problem, for planning auto-recovery of the IS state, and for executing the solution of the auto-recovery process of the IS state.

One of the possible algorithms for the functioning of the AMT (n)-automaton for auto-recovery task assignment with the predefined composition and OPC of its elastic tapes is proposed:

1. L_1 – ET user specifications (input tape)

$$\coprod_1 = \{P_1^1, P_{+1}^1, \alpha_1(*), \alpha_1(,), \alpha_1(\rightarrow)\}, F \overrightarrow{\prod} \left(\coprod_1 \right)$$
$$= \{\{S\} \rightarrow \{S\},\}^* \, a_1(*)a_1(\rightarrow)a_1(,) \tag{1.11}$$

2. L_2 = ET target installations (output tape)

$$\coprod_2 = \{P_1^2, P_{+1}^2, P_{-1}^2, P_a^2, W_a^2, I_{+1}^2 \, a_2(*), \alpha_2(,), \alpha_2(\rightarrow)\}, F \overrightarrow{\prod} \left(\sum\nolimits_2 \right)$$
$$= \{\{S\} \rightarrow \{S\},\}^* \, a_2(*)a_2(\rightarrow)a_2(,) \tag{1.12}$$

3. L_3 – ET system of grammar rules

$$\mathrm{II}_3 = \{P_1^3, P_{+1}^3, P_a^3, \alpha_3(*), \alpha_3(,), \alpha_3(\rightarrow)\}, F\,\vec{P}\left(\sum_3\right)$$
$$= \{\{S\} \rightarrow \{S\},\}^*\, a_2(*)a_2(\rightarrow)a_2(,) \tag{1.13}$$

4. L_4 – ET vocabulary

$$\mathrm{II}_4 = \{P_1^4, P_{+1}^4, P_a^4, \alpha_4(*), \alpha_4(,)\}, F\left(\sum_4\right) = \{S,\}^*\, a_4(*) \tag{1.14}$$

5. L_5 – ET (magazine)

$$\mathrm{II}_5 = \{P_1^5, P_{+1}^5, P_a^5, W_a^5, I_{+1}^5, I_{-1}^5, \alpha_5(*),\}, F\,\vec{P}\left(\sum_5\right) = \{S,\}^*\, a_5(*) \tag{1.15}$$

Thus, the technology of synthesis of abstract auto-recovery programs consists of the design of appropriate means of auto-projection of functional structures, the development of data structures and algorithms for the information transformation, and their subsequent abstract implementation. The design involves the specification preparation for an automated auto-recovery process, the structural decomposition of the process to the level of the basic models (the task assignment, the solution planning, the solution implementation), and decomposition of the basic models to the level of the scheme types (data, technologies, procedures). The result of the design is an informal specification of the technological process of self-recovery and a list of typical schemes. The development of algorithms and data structures involves the identification of schemes and the determination of the possible application of *PC* and *OPC* standard, the design of data circuits, knowledge, technologies, and procedures based on standard constructs, *PC* and *OPC* layout, and the assembly of *C*-grammar productions. At the implementation stage, the *ATP (n)* automata are synthesized according to the *C*-grammar, which consists in bringing the *PC* to a superposition of elementary operators and conditions, filling the information component with the *OPC* obtained earlier. The resulting system of models that provides information representation in the *OPC* and procedures for manipulating them in the PC and the corresponding base block for the interpretation of information schemes form the basis of the synthesis system of abstract auto-recovery programs for disturbed computations.

Thus, an abstract semantically controlled translator from the language of disturbed calculations into a verified plan for their recovery was developed. The types of destructive influences of intruders are interpreted, and the most realistic model of natural disturbances has been chosen. A structured abstract auto-recovery program to obtain a neutralization plan for the simulated computational disturbance was synthesized. The correspondence between the neutralization plan of the simulated disturbed calculation and the functional specification of the protected computation has been verified. A verified plan of disturbed computation neutralization under the real recovery procedures has been generated.

Based on the obtained results, a methodology for IS auto-recovery in conditions of destructive cyber-attacks using the system of immunities was developed and presented [217, 222, 244].

In general, the following new scientific and practical results were obtained.

In the analysis of approaches to the problem of IS auto-recovery under cyber-attacks:

- Methods of computation organization in software and technical influences were systematically analyzed, and the stability requirements of IS state operation were determined. High vulnerability and insufficient stability of IS functioning under the destructive influences are shown. The inconsistency of known information protection methods was demonstrated, and methods were developed for the control and recovery of computing processes to organize stable IS functioning under the destructive impacts of intruders.
- The fact was established that solving the problem of IS auto-recovery in the mass and group nature of cyber-attacks requires the search for new ways of organizing resilient IS functioning.
- The expediency was justified of developing models and methods of stable computation organization undergoing cyber-attacks by means that help the IS to develop immunity to disturbances of computational processes under mass attacks, like the immune protection system of a living organism and the ability to resist a cyber-attack in real time.

As for the concept development of ensuring the IS behavior resilience based on auto-recovery:

- Concepts of computations with memory for countering cyber-attacks and auto-recovery of disturbed IS states are introduced.
- Structuring principles of representation models of disturbed IS states and the corresponding stratification of the organization system of stable system behavior undergoing cyber-attacks are proposed.
- Scientifically methodical device, suitable for the problem solution of the organization of resilient IS functioning, is chosen and proved:

  - Model of the abstract computer of IS auto-recovery during cyber-attacks based on the dynamic system theory
  - Organization architecture of resilient IS functioning based on the theory of multi-level hierarchical systems
  - Language description of previously unknown types of structures of mass disturbances based on the theory of formal languages and grammars
  - Self-applicable translational model of abstract program synthesis of self-recovery of disturbed IS states for mass and group disturbances based on the methods of theoretical and system programming

- Architecture of the technological environment for the synthesis of abstract auto-recovery programs
- Specification of the behavior dynamics of disturbed calculations by analogy with disturbance modeling in living nature based on the catastrophe theory
- Methods of computability feature proof of reconstructed computations based on the similarity theory
- Method of immunity formation to destructive disturbances with result application of the computation control and recovery theory

As for the development of a scientific and methodical apparatus for the organization of the resilient IS operation in cyber-attacks:

- Model of IS auto-recovery, resistant to destabilization in a hierarchical multi-level control environment with feedback coupling, is developed. The model opens the essential possibility to study the IS functioning under cyber-attacks.
- Models of the simplest disturbances are designed to synthesize the scenarios of the process return of the IS functioning to equilibrium using the dynamic equations of the catastrophe theory. The methods allowed revealing quantitative patterns of counteraction to cyber-attacks and to synthesize a macro program of IS self-recovery in group and mass cyber-attacks for the first time.
- Representation models of mutually complementary denotational, logical, and operational semantics of correct IS states are proposed, making it possible to study the static and dynamic invariants of the system behavior at first and then to develop a micro program of auto-recovery operations and to prove its correspondence to the macro program of the auto-recovery scenario.
- Method of generalized multi-model verification of auto-recovery programs is proposed.
- Methods of IS auto-recovery with a use of the permissive standards, allowing to develop operational procedures for auto-recovery, are developed.

In the field of methodological support of the IS self-recovery organization:

- The system of development and accumulation of immunities was designed to provide comprehensive support to the organization of resilient IS functioning in cyber-attacks, including:

  - Application project, models, and technology of a self-applicable translator for the synthesis of abstract programs for IS auto-recovery under disturbances
  - Representation system and language of program-technical knowledge of IS self-recovery by regular schemes of algorithms
  - Model of abstract memory of AMT automata with extended signature of objects and operations
  - Operational automata of the translator knowledge system of abstract auto-recovery programs
  - Automatic model of the processor implementation of the regular schemes for IS auto-recovery

- A technology of prevention, detection, and neutralizing the mass cyber-attacks in the immunity ideology is proposed, based on:
  - Models and methods of automated monitoring of cyber-attacks and accumulation of immunity
  - Models of intruders in types of destructive impacts
  - Models representing the disturbance dynamics of the IS behavior and determining the scenarios for the return of the IS state to an equilibrium (stable) state
  - Macro models (programs) of IS auto-recovery under mass and group disturbances
  - Micro models (programs) of IS auto-recovery under mass and group disturbances
  - Models of denotational, axiomatic, and operational semantics of disturbed IS states
  - Method of proving the partial correctness of the IS features in auto-recovery
  - Method of operational standard output for the IS recovery
  - Presentation models of IS recovery
  - Development and implementation procedures of the IS recovery plan
- A technique for detection and neutralization of information and technical impacts on IS by an immunity system was developed.
- Methodical and practical recommendations on the IS design with the immunity system to mass cyber-attacks of the attacker are offered.

References

1. The Information Security Doctrine of the Russian Federation (approved by the Decree of the President of the Russian Federation No. 646 of December 5, 2016).
2. About formal bases of OWL [Electronic resource]. Access mode: http://semanticfuture.net/index.php. Accessed 20 Dec 2014
3. Abramov, S.M.: Research in the field of supercomputer technologies of the IPS RAS: a retrospective and perspective. In: Proceedings of the International Conference "Software Systems: Theory and Applications", vol. 1, pp. 153–192. Publishing house "University of Pereslavl", Pereslavl (2009)
4. Abramov, S.M.: History of development and implementation of a series of Russian supercomputers with cluster architecture. In: History of Domestic Electronic Computers. 2nd edn, Rev. and additional; color. Ill.: Publishing house "Capital Encyclopedia", Moscow (2016)
5. Abramov, S.M., Lilitko, E.P.: State and prospects of ultra-high performance computing systems development. Inf. Technol. Comput. Syst. 2, 6–22 (2013)
6. Action plan. Document WSIS-03/GENEVA/DOC/5-R dated December 12, 2013. Geneva [Electronic resource]. Access mode: http://www.itu.int/dms_pub/itus/md/03/wsis/doc/S03-WSIS-DOC-0005*PDF-R.pdf
7. Active Engagement, Modern Defence. Strategic Concept for the Defence and Security of the Members of the North Atlantic Treaty Organisation adopted by Heads of State and Government in Lisbon. November 19, 2010 [Electronic resource]. Access mode: http://www.nato.int/cps/en/SID-14EF0623-198FC77E/natolive/official_texts_68580.htm

8. Administration Strategy On Mitigating The Theft Of U.S. Trade Secrets. Executive Office of the President of the United States. February 2013, Washington, DC [Electronic resource]. Access mode: http://www.whitehouse.gov/sites/default/files/omb/IPEC/admin_strategy_on_mitigating_the_theft_of_u.s._trade_secrets.pdf

9. Advances in the field of information and telecommunications in the context of international security. Report of the UN Secretary-General. Document A/66/152 of 15 July 2011 [Electronic resource]. Access mode: http://www.un.org/en/documents/ods.asp?m=A/66/152

10. Advances in the field of information and telecommunications in the context of international security. Report of the First Committee. Document A/66/407 dated November 10, 2011 [Electronic resource]. Access mode: http://www.un.org/en/documents/ods.asp?m=A/66/407

11. Advances in the field of information and telecommunications in the context of international security. Resolution of the General Assembly of the UN. Document A/RES/65/41 dated December 8, 2010 [Electronic resource]. Access mode: http://www.un.org/en/documents/ods.asp?m=A/RES/65/41

12. Advances in the field of information and telecommunications in the context of international security. Resolution of the General Assembly of the UN. Document A/RES/68/243 dated December 27, 2013 [Electronic resource]. Access mode: http://www.un.org/en/ga/search/view_doc.asp?symbol=A/RES/68/243

13. Advancing America's Networking and Information Technology Research and Development Act of 2013. H. R. 967 [Electronic resource]. Access mode: https://www.govtrack.us/congress/bills/113/hr967/text

14. Agreement between the governments of the member states of the Shanghai Cooperation Organization on cooperation in the field of international information security from June 16, 2009, Yekaterinburg. Appendix 1. [Electronic resource]. Access mode: https://ccdcoe.org/sites/default/files/documents/SCO-090616-IISAgreementRussian.pdf

15. Aldrich, R.W.: The International Legal Implications of Information Warfare [Electronic resource]. Airpower J. **10**(3), 99–110 (1996). Access mode: http://www.airpower.maxwell.af.mil/airchronicles/apj/apj96/fall96/aldrich.pdf

16. Alekseeva, I.Y., et al.: Information Challenges of National and International Security; [under the Society. ed. A. V. Fedorova, VN Tsigichko], 328 p. PIR Center, Moscow (2001)

17. Alessandri, D., et al.: Towards a Taxonomy of Intrusion-Detection Systems and Attacks. Zurich, IBM Research Division (2001)

18. Almgren, M.: Consolidation and evaluation of IDS taxonomies. In: Proceedings of the Eight Nordic Workshop on Secure IT Systems, NordSec 2003

19. An evaluation Framework for National Cyber Security Strategies [Electronic resource]. European Union Agency for Network and Information Security (2014). Access mode: https://www.enisa.europa.eu/activities/Resilienceand-CIIP/national-cyber-security-strategies-ncsss/an-evaluation-framework-for-cyber-security-strategies-1

20. An Open, Safe and Secure Cyberspace. Joint communication to the European Parliament, the Council, the European Economic and Social committee and the Committee of the Regions Cybersecurity Strategy of the European Union of the European Commission and Higher Representative for foreign affairs and security policy. Brussels (2013) [Electronic resource]. Access mode: http://ec.europa.eu/information_society/newsroom/

21. Anderson, J.P.: Computer Security Threat Monitoring and Surveillance. James P. Anderson Co., Fort Washington, PA (1980)

22. Andreev, V.V., Zdiruk, K.B.: IV Jupiter: implementation of corporate security policy in computer networks. Open. Syst. **7–8**, 43–46 (2003)

23. Annual Incident Reports 2014: Analysis of Article 13a annual incident reports / European Union Agency for Network and Information Security (ENISA) (2015). [Electronic resource]. Access mode: https://www.enisa.europa.eu/activities/Resilience-and-CIIP/Incidents-reporting/annual-reports/annual-incident-reports-2014. Accessed 10 Apr 2016

24. Appliance of information and communication technologies for development. Resolution of the General Assembly of the UN. Document A/RES/65/141 dated December 20, 2010 [Electronic resource]. Access mode: http://www.un.org/en/ga/search/view_doc.asp?symbol=A/RES/65/141

25. Arbatov A.G. Real and imaginary threats: Military power in world politics in the beginning of the XXI century. [Electronic resource] AG Arbatov. Russia in global politics. March 3, 2013. Access mode: http://www.global- affairs.ru/number/Ugrozy-realnye-i-mnimye-15863

26. Aristotle. Comp. in 4 volumes (Series "Philosophical heritage"). Thought, Moscow. (1975–1983)

27. Arquilla, J.: Ethics and information warfare. In: Khalilzad, Z., White, J., Marsall, A. (eds.) Strategic Appraisal: The Changing Role of Information in Warfare, 475 p. RAND Corporation, Santa Monica (1999)

28. Ashby, U.R.: Principles of Self-Organization, pp. 314–343. Mir, Moscow (1966)

29. Axelsson, S.: Intrusion Detection Systems: A Taxonomy and Survey. Technical Report 99–15. Department of Computer Engineering, Chalmers University of Technology, Goteborg (2000)

30. Barabanov, A.V., Markov, A.S., Tsirlov, V.L.: Methodological framework for analysis and synthesis of a set of secure software development controls. J. Theor. Appl. Info. Technol. **88** (1), 77–88 (2016)

31. Barabanov, A., Lavrov, A., Markov, A., Polotnyanschikov, I., Tsirlov, V.: The study into cross-site request forgery attacks within the framework of analysis of software vulnerabilities. In: Preliminary proceedings of the 11th Spring/Summer Young Researchers' Colloquium on Software Engineering (Innopolis, Republic of Tatarstan, Russian Federation, June 5–7, 2017), pp. 105–109. SYRCoSE, ISP RAS

32. Baranov, P.A.: Detection of anomalies based on the application of the criterion of the dispersion degree. Proceedings of the XIV All-Russian Scientific Conference "Information Security Problems in the Higher School System", pp. 25–27. Izd. department of the St. Petersburg State Polytechnic University, St. Petersburg (2007)

33. Batueva, E.V.: American concept of threats to information security and its international political component, 207 p. Doctoral thesis of political sciences. MGIMO (U) Ministry of Foreign Affairs of the Russian Federation, Moscow (2014)

34. Bedritsky, A.V.: American policy of cyber space control. Probl. Natl. Strat. **2**(3), 25–40 (2010)

35. Bedritsky, A.V.: Information War: Concepts and Their Implementation in the US, 183p. RISI, Moscow (2008)

36. Bedritsky, A.V.: The Evolution of the American Concept of Information War, 26p. RISI, Moscow. Analytical Rev. (3) (2003)

37. Biryukov, D.N.: Cognitive-functional memory specification for simulation of purposeful behavior of cyber systems. Proc. SPIIRAS. **3**(40), 55–76 (2015)

38. Biryukov, D.N., Lomako, A.G.: Denotational semantics of knowledge contexts in ontological modeling of the subject areas of conflict. Proc. SPIIRAS. **5**(42), 155–179 (2015)

39. Biryukov, D.N., Glukhov, A.P., Pilkevich, S.V., Sabirov, T.R.: Approach to the processing of knowledge in the memory of an intellectual system. Natur. Tech. Sci. **11**, 455–466 (2015)

40. Biryukov, D.N., Lomako, A.G.: Approach to the construction of information security systems capable of synthesizing scenarios of anticipatory behavior in the information conflict. Protect. Inf. Inside. **6**(60), 42–50 (2014)

41. Biryukov, D.N., Lomako, A.G.: The formalization of semantics for representation of knowledge about the behavior of conflicting parties: materials of the 22nd scientific-practical conference "Methods and technical means of information security", pp. 8–11. Publishing house of Polytechnic University, St. Petersburg (2013)

42. Biryukov, D.N., Lomako, A.G., Petrenko, S.A.: Generating scenarios for preventing cyberattacks. Protect. Inf. Inside. **4**(76) (2017)

43. Biryukov, D.N., Lomako, A.G., Rostovtsev, Y.G.: The appearance of anticipatory systems to prevent the risks of cyber threat realization. Proc. SPIIRAS. **2**(39), 5–25 (2015)

44. Biryukov, D.N., Lomako, A.G., Sabirov, T.R.: Multilevel Modeling of Pre-Emptive Behavior Scenarios. Problems of Information Security. Computer systems, vol. 4, pp. 41–50. Publishing house of Polytechnic University, St. Petersburg (2014)
45. Biryukov, D.N., Rostovtsev, Y.G.: Approach to constructing a consistent theory of synthesis of scenarios of anticipatory behavior in a conflict. Proc. SPIIRAS. **1**(38), 94–111 (2015)
46. Biryukov, D.N., Lomako, A.G.: Approach to Building a Cyber Threat Prevention System. Problems of Information Security. Computer systems, vol. 2, pp. 13–19. Publishing house of Polytechnic University, St. Petersburg (2013)
47. Bocharov, V.A., Markin, V.I.: Fundamentals of Logic. Moscow State University, Moscow (2008)
48. Boev, S.F., Kochkarov, A.A., Stupin, D.D.: Development of R & D activities of high-tech B2G-holdings: problems and tasks. Qual. Innov. Educ. **11**(78), 54–59 (2011)
49. Boev, S.F., Kochkarov, A.A., Stupin, D.D.: The role and possibilities of pre-university training in the problem of the formation of highly qualified specialists for high-tech branches of the real economy and the experience of the RTI Systems Concern: materials of the International Scientific Conference "Forming the Identity of Finno-Ugric world and Russian education", pp. 330–333. Mordovian state publishing house University, Saransk (2011)
50. Bongard, M.M.: The Problem of Recognition. Fizmatgiz, Moscow (1967)
51. Brennen, S.: Cyberthreats and the Decline of the Nation-state, 175 p. Susan W. Brenner. Routledge, Abingdon (2014)
52. Brenner, J.: America the Vulnerable, 308 p. Joel Brenner. Penguin Press, New York (2011)
53. Carr, J.: Inside Cyber Warfare, 213 p. Jeffrey Carr. O'Reilly (2010)
54. Cavelty, M.: Cyber-Security and Threat Politics: US Efforts to Secure the Information Age, 182 p. Myriam Dunn Cavelty. Routledge, New York (2007)
55. Chereshkin, D.S.: Problems of Information Security Management, 224 p. Editorial URSS, Moscow (2002)
56. Clarifying Cybersecurity Responsibilities and Activities of the Executive Office of the President and the Department of Homeland Security. Memorandum. Executive Office of the President Office of Management and Budget, Washington, DC. July 6, 2010 [Electronic resource]. Access mode: http://www.whitehouse.gov/sites/default/files/omb/assets/memoranda_2010/m10-28.pdf
57. Clark, R., Nake, R.: The Third World War. What Will It Be Like? Publishing house "Peter", St. Petersburg (2011)
58. Clark, W., Levin, P.: Securing the information highway: How to enhance the United States electronic defenses. Foreign Aff. November/December 2009 [Electronic resource]. Access mode: http://www.foreignaffairs.com/articles/65499/wesley-k-clark-and-peter-llevin/securing-the-information-highway
59. Clarke, R.: Cyber War the Next Threat to National Security and What to Do About It. In: Richard A. Clarke, Robert K. Knake, 290 p. HarperCollins (2010)
60. Clarke, R.: Securing Cyberspace Through International Norms. Good Harbor Security Risk Management [Electronic resource]. Access mode: http://www.goodharbor.net/media/pdfs/SecuringCyberspace_web.pdf
61. Clayton, M.: Presidential Cyberwar Directive Gives Pentagon Long-awaited Marching Orders. The Christian Science Monitor. June 10, 2013 – [Electronic resource]. Access mode: http://www.csmonitor.com/USA/Military/2013/0610/Presidential-cyberwardirective-gives-Pentagon-long-awaited-marching-orders-video
62. Collin, B.: The Future of Cyberterrorism. Crime Justice Int. **13**(2) March 1997 [Electronic resource]. Access mode: http://www.cjimagazine.com/archives/cji4c18.html?id=415
63. Collins, A.M., Quillian, M.R.: Retrieval time from semantic memory. J. Verbal Learn. Verbal Behav. **8**, 240–247 (1969)
64. Communication from the Commission to the European Parliament and the Council. The EU Internal Security Strategy in Action: Five steps towards a more secure Europe. Brussels, 22.11.2010. COM (2010)

65. Comprehensive National Cybersecurity Initiative. The White House, Washington, DC. January 2008 [Electronic resource]. Access mode: http://www.whitehouse.gov/cybersecurity/compre hensive-nationalcybersecurity-initiative

66. Consolidated and Further Continuing Appropriations Act of 2013. H. R. 933 [Electronic resource]. Access mode: http://www.gpo.gov/fdsys/pkg/BILLS-113hr933pp/pdf/BILLS-113hr933pp.pdf

67. Cornish, P.: Cyber security and politically, socially and religiously motivated cyber-attacks. 2009 [Electronic resource]. Access mode: http://www.europarl.europa.eu/activities/commit tees/studies.do?language=EN

68. Creation of a global culture of cybersecurity and assess national efforts to protect critical information infrastructures. UN Resolution. Document A/RES/64/211 dated December 21, 2009 [Electronic resource]. Access mode: http://www.un.org/en/documents/ods.asp? m=A/RES/64/211

69. Crimes involving the use of a computer network. The Tenth United Nations Congress on the Prevention of Crime and the Treatment of Offenders. Document A / CONF.187 / 10 of 3 February 1999

70. Critical Infrastructure Research and Development Advancement Act of 2013. H. R. 2952 [Electronic resource]. Access mode: https://www.govtrack.us/congress/bills/113/hr2952/text

71. Critical Infrastructure Security and Resilience: Presidential Policy Directive/PPD-21. The White House, Washington, DC. February 12, 2013

72. Cyber Europe 2012: Key Findings Report. ENISA. 2012 [Electronic resource]. Access mode: https://www.enisa.europa.eu/activities/Resilience-and-CIIP/cyber-crisis-cooperation/cce/ cyber-europe/cyber-europe-2012/cyber-europe2012-key-findings-report. Accessed date 10 Apr 2016

73. Cyber Intelligence Sharing and Protection Act. 2012. H. R. 3523 [Electronic resource]. Access mode: https://www.govtrack.us/congress/bills/112/hr3523

74. Cyber Security Report. European Commission. 2013 [Electronic resource]. Access mode: http://ec.europa.eu/public_opinion/archives/ebs/ebs_404_en.pdf. Accessed date 10 Apr 2016

75. Cyber Security Report. European Commission. 2015. [Electronic resource]. Access mode: http://ec.europa.eu/COMMFrontOffice/PublicOpinion/index.cfm/Survey/getSurveyDetail/ yearFrom/1973/yearTo/2016/search/cyber/surveyKy/2019. Accessed 10 Apr 2016

76. Cyberpower and National Security [ed. F. Kramer, S. Starr, and L. Wentz], 664 p. Potomac Books Inc. (2009)

77. Cybersecurity Act of 2009. S.773. Open Congress Summary [Electronic resource]. Access mode: http://www.opencongress.org/bill/111-s773/show

78. Cybersecurity Strategy of the European Union: An Open, Safe and Secure Cyberspace. High Representative of the European Union for Foreign Affairs and Security Policy. Brussel, 2013 [Electronic resource]. Access mode: http://eeas.europa.eu/policies/eu-cyber-security/cybsec_ comm_en.pdf. Accessed 10 Apr 2016

79. Cyberspace Policy Review Assuring a Trusted and Resilient Information and Communications Infrastructure. May 2009 [Electronic resource]. Access mode: http://www.whitehouse.gov/ assets/documents/Cyberspace_Policy_Review_final.pdf

80. Debar H., et al.: (IBM Zurich). Towards a Taxonomy of Intrusion-Detection Systems. IBM Research Division, Zurich (1999)

81. Decree of the Government of the Russian Federation of 04 September 2003 No. 547 "On the preparation of the population in the field of protection from natural and man-made emergency situations"

82. Decree of the Government of the Russian Federation of December 30, 2003 No. 794 "On Unified State System for the Prevention and Elimination of Emergency Situations"

83. Decree of the Government of the Russian Federation of December 8, 2011 No. 2227-r "On the Approval of the Strategy for Innovative Development of the Russian Federation for the Period to 2020". [Electronic resource]. Access mode: http://mon.gov.ru/files/materials/4432/11.12. 08-2227r.pdf. 145. RD 50-34.698-90. Automated systems. Requirements for the content of documents

84. Denning, D.: Cyberterrorism. George Town University. May 23, 2000 [Electronic resource]. Access mode: http://www.cs.georgetown.edu/~denning/infosec/cyberterror.html
85. Denning, D.: Information Operations and Terrorism / Defense Technical Information Center. August 18, 2005. [Electronic resource]. Access mode: http://www.dtic.mil/cgi-bin/ GetTRDoc?AD=ADA484999
86. Denning, D.: Information Warfare and Security, 522 p. ACM Press, New York (1999)
87. Denning, D.: Is cyberterror next? Social Science Research Council. November 1, 2001 [Electronic resource]. Access mode: http://essays.ssrc.org/sept11/essays/denning.htm
88. Denning, D.: Reflections on cyberweapons controls. Comput. Security J. **XVI**(4), 43–53 (2000)
89. Denning, D.E., (SRI International): An intrusion detection model. IEEE Trans. Softw. Eng. (SE-13), **2**, 222–232 (1987)
90. Department of Defense Dictionary of Military and Associated Terms. November 8, 2010 [Electronic resource]. Access mode: http://www.dtic.mil/doctrine/dod_dictionary/
91. Department of Defense Strategy for Operating in Cyberspace. July 2011. [Electronic resource]. Access mode: http://www.defense.gov/news/d20110714cyber.pdf
92. Digital Agenda for Europe. A Europe 2020 Strategy. 2010 [Electronic resource]. Access mode: http://ec.europa.eu/digitalagenda.
93. Dunlap, C. Jr.: Perspectives for cyber strategists on law for cyberwar (Charles J. Dunlap Jr.). Strateg. Stud. Q. Spring, 81–99 (2011)
94. Electronic Communications Privacy Act Amendments Act of 2013. S. 607 [Electronic resource]. Access mode: https://www.govtrack.us/congress/bills/113/s607
95. Elliott, D.: Weighing the Case for a Convention to Limit Cyberwarfare. Arms Control Association. November 2009 [Electronic resource]. Access mode: http://www.armscontrol. org/act/2009_11/Elliott
96. Ermakov, S.M.: Transformation of NATO after the Lisbon Summit in 2010: from the defense of the territory to the protection of the public domain. Probl. Natl. Strateg. **4**(9), 107–128 (2011)
97. Terrorist Use of the Internet: Information Operations in Cyberspace. Congressional Research Service. March 8, 2011. 16 p. [Electronic resource]. Access mode: http://www.fas.org/sgp/crs/ terror/R41674.pdf
98. The concept of foreign policy of the Russian Federation (approved by the Decree of the President of the Russian Federation of November 30, 2016 No. 640
99. The concept of the development of an intelligent electric power system in Russia with an actively adaptive network. OJSC "FGC UES" OJSC "Scientific and technological center of electric power industry". Moscow (2011)
100. The concept of the state system for detecting, preventing and eliminating the consequences of computer attacks on the information resources of the Russian Federation (approved by the President of the Russian Federation on December 12, 2014, No. K 1274)
101. The Economic Impact of Cybercrime and Cyber Espionage. The Center for Strategic and International Studies Report. July 2013. 19 p. [Electronic resource]. Access mode: http://csis. org/files/publication/60396rpt_cybercrimecost_0713_ph4_0.pdf
102. The European Cyber Security Month 2015: Deployment report. European Union Agency for Network and Information Security (ENISA). 2015 [Electronic resource]. Access mode: https:// www.enisa.europa.eu/activities/stakeholder-relations/nis-brokerage-1/european-cyber-secu rity-month-advocacy-campaign/2015. Accessed 10 Apr 2016
103. The national security strategy of the Russian Federation (approved by the Decree of the President of the Russian Federation of December 31, 2015, No. 683
104. The Order of the Ministry of Emergency Measures of the Russian Federation from February, 28th, 2003 ⬚ 105. On the statement of requirements on the prevention of extreme situations on potentially dangerous objects and objects of life-support
105. The Regulation on Cooperation of the Member States of the Collective Security Treaty Organization in the Sphere of Ensuring Information Security of December 10, 2010. [Electronic resource]. Access mode: http://docs.pravo.ru/document/view/16657605/14110649/. 129. Pospelov DA Thinking and automatons, 224 p. Soviet radio, Moscow (1972)

106. The role of science and technology in the context of international security, disarmament and other related fields. Report of the First Committee. Document A/53/576 of 18 November 1998 [Electronic resource]. Access mode: http://www.un.org/en/documents/ods.asp?m=A/53/576

107. The Russia U.S. Bilateral on Cybersecurity – Critical Terminology Foundations. EastWest Institute. Issue 1. April 2011. 47 p. [Electronic resource]. Access mode: http://www.ewi.info/idea/russia-us-bilateral-cybersecurity-criticalterminologyfoundations

108. The Stuxnet Computer Worm: Harbinger of an Emerging Warfare Capability. Congressional Research Service. December 9, 2010. 9 p. [Electronic resource]. Access mode: http://fas.org/sgp/crs/natsec/R41524.pdf

109. Thomas, T.: Cyber Silhouettes. Shadows Over Information Operations, 334 p. Timothy L. Thomas. Foreign Military Studies Office (FMSO). Fort Leavenworth (2005)

110. Thomas, T.: Is the IW paradigm outdated? A discussion of U.S. IW theory. J. Inf. Warfare. 2 (3), 109–116 (2003)

111. Threats Posed by the Internet. Threat Working Group of the CSIS Commission on Cybersecurity for the 44th Presidency. October 2008. 28 p. [Electronic resource]. Access mode: http://csis.org/files/media/csis/pubs/081028_threats_working_group.pdf

112. Toffler, A.: War and Anti-War: Survival at the Down of the Twenty-First Century, 1st edn, 302 p. Alvin and Heidi Toffler (1993)

113. Toffler, E.: The Third Wave, 784 p. AST, Moscow (2010)

114. Tsygichko, V.N., Votrin, D.S., Krutskikh, A.V., Smolyan, G.L., Chereshkin, D.S.: Information Weapons Are a New Challenge to International Security, 52 p. Institute of System Analysis of the Russian Academy of Sciences, Moscow (2000)

115. Tulving E. Episodic and Semantic Memory. Organization of Memory New York: Academic, 1972. P. 381–403.

116. Unsecured Economies: Protecting Vital Information. McAfee Report. (2009) 33 p. [Electronic resource]. Access mode: https://resources2.secureforms.mcafee.com/LP=2984

117. Vasyutin, S.V., Zavyalov, S.S.: Neural network method for analyzing the sequence of system calls for the detection of computer attacks and the classification of application modes. Methods and Means of Information Processing: Proceedings of the Second All-Russian Scientific Conference; [ed. member corr. RAS L.N. Koroleva], pp. 142–147. Pub. Department of the Factor of Computational Mathematics and Cybernetics of the Moscow State University. M.V. Lomonosov, Moscow (2005)

118. Wales Summit Declaration. Issued by the Heads of State and Government participating in the meeting of the North Atlantic Council in Wales. September 5, 2014 [Electronic resource]. Access mode: http://www.nato.int/cps/en/natohq/official_texts_112964.htm

119. Weimann, G.: Cyberterrorism. How Real Is the Threat? United States Institute of Peace. Special Report. 12 p. [Electronic resource]. Access mode: http://www.usip.org/sites/default/files/sr119.pdf

120. Weimann, G.: Special Report 116: www.terror.net How Modern Terrorism Uses the Internet/ United Institute of Peace, March 2004. [Electronic resource]. Access mode: http://dspace.cigilibrary.org/jspui/bitstream/123456789/4610/1/www%20terror%20net%20How%20Modern%20Terrorism%20Uses %20the%20Internet.pdf?

121. Petrenko, A.A., Petrenko, S.A.: Cyber units: methodical recommendations of ENISA. Quest. Cybersecurity. 3(11), 2–14 (2015)

122. Petrenko, A.A., Petrenko, S.A.: The way to increase the stability of LTE-network in the conditions of destructive cyber-attacks. Quest. Cybersecurity. 2(10), 36–42 (2015)

123. Petrenko, S.A.: Methods of ensuring the stability of the functioning of cyber systems under conditions of destructive effects. Proceedings of the ISA RAS. Risk Manag. Security, 52, 106–151 (2010)

124. Petrenko, S.A., Kurbatov, V.A., Bugaev, I.A., Petrenko, A.S.: Cognitive system of early warning about computer attack. Protect. Inf. Inside. 3(69), 74–82 (2016)

125. Tallinn Manual on the International Law Applicable to Cyber Warfare. [Electronic resource] general editor Michael N. Schmitt. Cambridge University Press (2013). 282 p. Access mode: http://issuu.com/nato_ccd_coe/docs/tallinnmanual?mode=embed&layout=http%3A%2F%2Fskin.issuu.com%2Fv%2Flight%2Flayout.xml&showFlipBtn=true
126. Gamayunov, D.Y.: Detection of computer attacks based on the analysis of the behavior of network objects: dis. for the competition uch. degree of Cand. fiz.-mat. sciences. Moscow State University, Moscow (2007)
127. Petrenko, S.A.: Methods of detecting intrusions and anomalies of the functioning of cyber system, Proceedings of ISA RAS. Risk Manag. Safety. **41**, 194–202 (2009)
128. Ilgun, K.: USTAT: A real-Time Intrusion Detection System for UNIX. Computer Science Department, University of California, Santa Barbara (1992)
129. Kumar, S., Spafford, E.H.: An Application of Pattern Matching in Intrusion Detection. Purdue University, New York (1994)
130. Petrenko, A.S., Petrenko, S.A.: Designing of corporate segment SOPKA. Protect. Inf. Inside. **6** (72), 48–50 (2016)
131. Petrenko, S.A., Petrenko, A.S.: From detection to prevention: trends and prospects of development of situational centers in the Russian Federation. Intellect Technol. **1**(12), 68–71 (2017)
132. Petrenko, S.A., Shamsutdinov, T.I., Petrenko, A.S.: Scientific and technical problems of development of situational centers in the Russian Federation. Inf. Protect. Inside. **6**(72), 37–43 (2016)
133. Portnoy, L., et al.: Intrusion detection with unlabeled data using clustering. ACM Workshop on Data Mining Applied to Security (2001)
134. Kotenko, I.V.: Intellectual mechanisms of cybersecurity management. Proceedings of ISA RAS. Risk Manag. Safety, **41**, 74–103 (2009)
135. Petrenko, A.S., Petrenko, S.A.: Super-productive monitoring centers for security threats. Part 1. Protect. Inf. Inside. **2**(74), 29–36 (2017)
136. Petrenko, A.S., Bugaev, I.A., Petrenko, S.A.: Master data management system SOPKA. Inf. Protect. Inside. **5**(71), 37–43 (2016)

Chapter 2
Finite Capabilities of Cybersecurity Technologies

2.1 CERT/SCIRT Capacity Limits

In November 1988, the first computer security response center, CERT (*Computer Emergency Response Team*), was established at Carnegie Mellon University in Pittsburgh, Pennsylvania, USA. At present, around 300 CERT/CSIRT centers around the world have been established at various commercial, state, and educational organizations. These centers' relevance owes to the necessity for timely and high-quality delivery of professional services to diverse state and commercial organizations for the prevention, detection, and response to cybersecurity incidents. This chapter will give the evolutionary development of CERTs from local to global solutions, implying the construction of an extensive national network of such centers while simultaneously clarifying the peculiarities of creating a promising "cloud" center for responding to security incidents within the national state program "Information-Oriented Society (2011–2020)."

2.1.1 State-of-the-Art Review

The abbreviation *CERT/CSIRT* designates a special team (or group) which responds to cybersecurity incidents (e.g., *Cyber Emergency Response Team*, the *Cyber Security Incident Response Team*, the *Computer Emergency Readiness Team*, etc.). In this context, a cybersecurity incident is understood to be one or a series of undesirable or unexpected events in an information security system that could compromise an organization's operations and information security. These events point to a committed, attempted, or probable threat to information security.

© Springer International Publishing AG, part of Springer Nature 2018
S. Petrenko, *Big Data Technologies for Monitoring of Computer Security: A Case Study of the Russian Federation*, https://doi.org/10.1007/978-3-319-79036-7_2

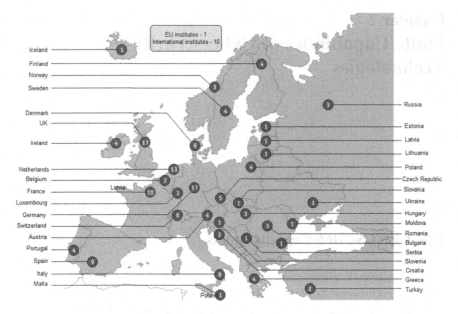

Fig. 2.1 Map of European CERT/CIRT (number of centers per country in May 2012)

The first CERT center was founded in the USA in November 1988 based at Carnegie Mellon University, after the malicious software, the "Morris worm," paralyzed critical Internet nodes of the US national infrastructure on November 3, 1988. To date, the center has become the leading nationwide coordinating center CERT/CC[1] with the right to officially distribute the corresponding trademark "CERT."

Most CERTs in the EU were founded by leading universities and major IT companies. Initially, the common interaction of these CERTs was carried out by the *TF-CSIRT (Task Force-Collaboration Security Incident Response Teams)*[2] and *FIRST (Forum of Incident Response and Security Teams)*[3] communities. At present, the management of the CERT centers of the EU is gradually being transferred to the European Network and Information Security Agency (ENISA).[4]

In Russia, several state and corporate computer security incident response centers have also been established (Fig. 2.1), with the most notable being COV-CERT.RU, CERT-GIB, RU-SERT, WebPlus ISP, CRSIT ARSIB, and other centers.

---

[1]http://www.cert.org/certcc.html/

[2]http://www.enisa.europa.eu/activities/cert/background/coop/status-quo/evaluation/tf-csirt/

[3]http://www.first.org/events/colloquia/lisbon2013/

[4]http://www.enisa.europa.eu/.

COV-CERT.RU[5] is the state computer incident response center involving objects located in the RSNET (*Russian State Network*) Internet segment, as well as other information infrastructure objects belonging to Russian state authorities.

The key functions of GOV-CERT.RU involve:

- Advisory and methodical assistance in conducting actions to eliminate the consequences of cyber incidents in Russian state ITS
- Analysis of incident causes and conditions in Russian state ITS
- Development of guidelines on current information security threat neutralization
- Collaboration with Russian, foreign, and international cyber incident response organizations
- Accumulation and analysis of cyber incident information

In general, the GOV-CERT.RU activities are aimed at responding to the following typical threats and incidents:

- "Denial-of-service" attacks against Russian state ITS
- Involvement of ITS objects in the botnet and the spread of malicious software
- Unauthorized access attempts to the Russian state ITS objects

RU-CERT[6] (*Russian Computer Emergency Response Team*) is an autonomous nonprofit organization called the "cyber incident response center." RU-CERT is part of the CSIRT/CERT international center associations (FIRST, *Trusted Introducer*).

The center primarily aims to reduce the level of information security threats for Internet users in the Russian segment. The center assists Russian and foreign legal entities and individuals in identifying, preventing, and suppressing illegal activities related to the network resources of the Russian Federation.

The center accumulates, stores, and processes statistical data on the propagation of malicious programs and network attacks in Russia. RU-CErT considers cyber incidents to be facts and signs of violation of Internet legislation at the Russian and international levels, as well as violations of the norms and rules of its use which most network users accept. At the same time, this center provides incident response services without any guarantees and liabilities.

RU-CERT cooperates with leading Russian IT companies, criminal investigation agencies, government authorities of the Russian Federation, foreign computer incident response centers, and other organizations that operate in the computer and information security domain. Acting within the regulatory legal framework of the Russian Federation, RU-CERT is not authorized to deal with issues managed by law enforcement agencies (*Russian FSB or the Russian Ministry of Internal Affairs*). Also, the center is not empowered to close resources, filter addresses, stop domain delegation, remove content from a particular resource, search for persons involved in certain actions, etc.

---

[5]http://gov-cert.ru

[6]http://cert.ru/ru/about.shtml

CERT-GIB[7] is a nongovernmental CERT in the Russian Federation territory and is located in the commercial company Group-IB. This center is authorized by the Carnegie Mellon University and has rights to use the trademark "CERT." In addition, the center is accredited by the international associations FIRST[8] and Trusted Introducer[9] and belongs to several cybersecurity alliances, one of which being the well-known IMPACT[10] supported by the United Nations (UN).

The CERT-GIB has the following essential tasks:

- Information exchange coordination on information security incidents (hereinafter IS incidents) between law enforcement authorities, legal entities, and individuals.
- Assistance with cybersecurity maintenance in the Russian segment of the Internet and beyond.
- Operational control assistance in managing cyber risks and eliminating sources of cyber threat. Within the framework of these tasks, CERT-GIB provides free 24-hour information support on security issues and IS incident response in RuNet.

The functions of the CERT-GIB also involve:

- Providing initial support and recommendations for DDoS incidents
- Informing legal entities and individuals about registered fraud cases
- Assisting injured parties in establishing contacts with registrars, hosting providers, and law enforcement agencies of the Russian Federation and the CIS
- Responding to fraudulent requests for site neutralization, such as phishing, fake banking sites, scam sites, sources of copyright infringement, and other types of malicious resources

Furthermore, CERT-GIB activities are aimed at responding to the following typical incidents:

- Denial-of-service attacks (DoS, DDoS)
- Unauthorized use of the data processing and storage systems
- Information compromising
- Asset compromising
- Internal/external unauthorized access
- Building and distribution of malicious software
- Information security policy violation
- Phishing and unlawful use of brand name on the Internet
- Fraudulent activities with RB systems and electronic payment systems

---

[7]http://cert-gib.ru

[8]http://www.first.org/members/teams/cert-gib

[9]http://www.trusted-introducer.org/directory/teams/cert-gib.html

[10]http://www.impact-alliance.org/download/pdf/media/whats-new/2013/ITU-IMPACT-Group-IB.pdf

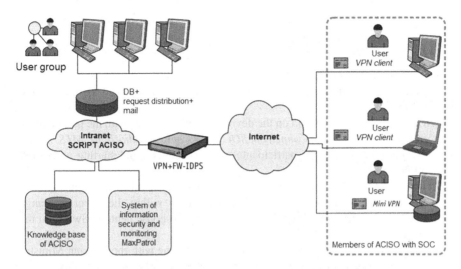

Fig. 2.2   Draft outline of CSIRT RACISO center

Furthermore, based on the agreement,[11] the scope of CERT-GIB competence includes counteracting the use of domain names for phishing, unauthorized access to third-party information systems, and the malicious software distribution and control (botnets).

RACISO CSIRT is a projected intersectoral center for responding to cybersecurity incidents based on the Russian Association of Chief Information Security Officer (RACISO).[12] The main task of this center is to assist state and corporate organizations to detect, prevent, and suppress security incidents in the Russian Federation (Fig. 2.2).

RACISO CSIRT construction is outlined in the following steps:

1. Choice of rooms for hardware deployment (server rack), as well as for personnel accommodation (stage 1 5X8).
2. Deployment of selected hardware for the center's infrastructure:

   - E-mail servers, FTP server (sftp), and server of knowledge base and queries
   - Work stations for the duty staff (domain)
   - RADIUS authentication system (perhaps, *RSA* tokens or *Aladdin* e-tokens)
   - Security systems for web surfing (*ASA*, *VPN-hub*, *IDPS*)

3. Deployment of selected hardware to provide services:

   - *MAXPATROL* audit systems
   - Certification authority system

---

[11]http://cert-gib.ru

[12]http://www.aciso.ru/structure/central_monitoring_csirt_ru.php

- Launched site for inspection of the threats for center subscribers with capacity to provide graphic data
- Systems to provide the security services

4. Writing of documentation for the center personnel.
5. Training for center employees to work with hardware, as well as studying the documents.
6. Creation of an active stand on the developed infrastructure with the participation of RACISO members (*Vympelcom SOC, CERT of Rosselkhozbank, MTS, LETA-IT*, and other concerned participants) with issuing IDs and starting work with VPN clients.
7. Improvement of the center's technical part and its test documentation.
8. Study the feasibility of exchanging information within the cyber exchange CYBEX X.1500 with other computer incident response centers. Switch to the stage 24X8.
9. Harmonization of the automated interaction scheme with the Russian Ministry of Internal Affairs and the Russian Federal Security Service on issues of cybersecurity incident response.
10. Accumulating a cybersecurity incident database and keeping it up to date.

2.1.2   Cloud Aspects of CERT/CSIRT

In 2012, a number of well-known international professional IT organizations began to work out the issues of creating a network of national CERTs based on cloud technologies. These include the *International Telecommunication Union (ITU)*, the *Institute of Electrical and Electronics Engineers (IEEE)*, the *International Organization for Standardization (ISO)*, the *European Network Agency And Information Security (ENISA)*, the *National Institute of Standards and Technology (NIST)*, CERT *Coordination Center (CERT/CC) CMU, Association for Computing Machinery*, the *World Wide Web Consortium (W3C)*, the *Cloud Security Alliance (CSA)*, the *Internet Security Alliance (ISA)*, and others (Figs. 2.3 and 2.4).

Among the abovementioned initiatives, the recommendations of the *International Telecommunication Union (ITU)* are highlighted.

ITU was founded in 1865 in Paris as the International Telegraph Union. In 1934, it received its modern name, and in 1947, it became a specialized agency of the United Nations. At present, ITU includes 193 countries and about 700 business institutions.

In addition to its headquarters in Geneva, 12 regional and area offices around the world[13] have been established.

---

[13]http://www.itu.int/ru/about/Pages/history.aspx

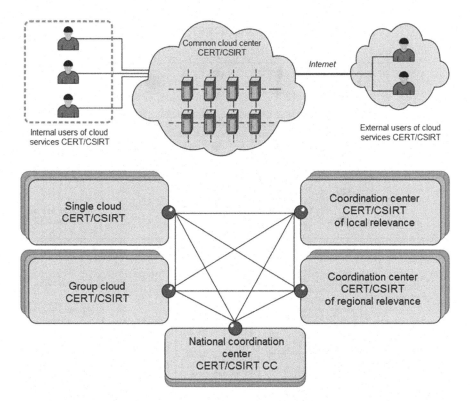

Fig. 2.3 Feasible physical architecture of the cloud CERT/CSIRT

The ITU functional institutions include:

- The *World Telecommunication Standardization Assembly*, held every 4 years, is the main decision-making body of the standardization sector.
- The *Telecommunication Standardization Bureau* is the executive branch of the standardization sector.
- Research groups.
- The *Telecommunication Standardization Advisory Group* is an auxiliary branch conducting coordination work.

The Union's supreme body is the Plenipotentiary Conference, a meeting of delegations of the member states of the Union, which is held every 4 years. The main executive bodies are the Council and the General Secretariat.

The main work branches are divided into three sectors:

- Telecommunication standardization sector, ITU-T
- Radiocommunication sector, ITU-R
- Telecommunication development sector, ITU-D

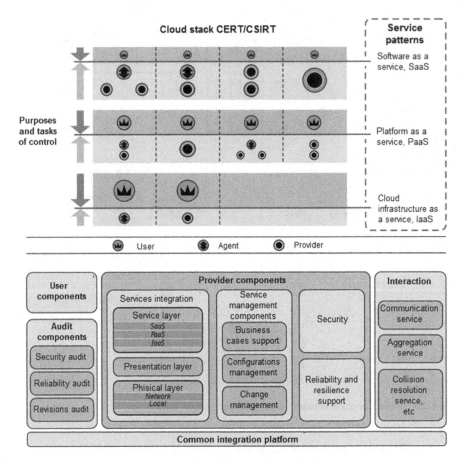

Fig. 2.4 Feasible logical architecture of the cloud CERT/CSIRT

In accordance with the resolution of the First World Telecommunication Standardization Assembly (2000), the practice was adopted of appointing *Lead Study Groups* (*LSGs*) on certain issues requiring simultaneous coordination of the efforts of several research groups working in various fields (Fig. 2.5).

In September 2001, Study Group 17: Data Networks and Telecommunication Software, formed on the basis of previous groups "Study Group 7" and "Study Group 10," began its work. Currently, Group 17 is the leading research group on security (Communication Systems Security, CSS) (Table 2.1).

The main activities of Group 17 involve:

- Network management security (recommendations: M.3010, principles of telecommunications management networks; M.3016, overview of telecommunication management networks security; etc.)

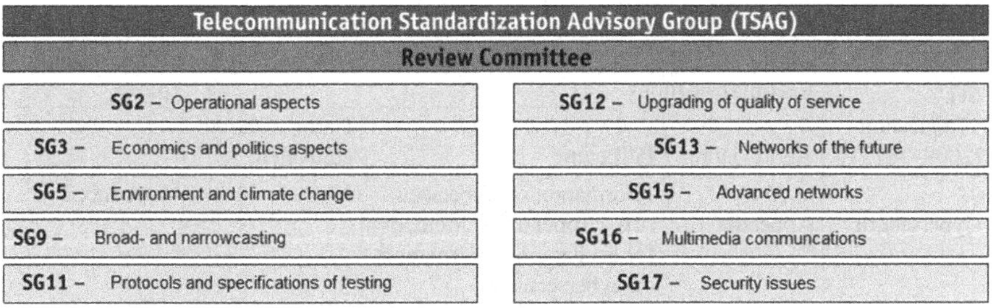

Telecommunication Standardization Advisory Group (TSAG)	
Review Committee	
SG2 – Operational aspects	**SG12** – Upgrading of quality of service
SG3 – Economics and politics aspects	**SG13** – Networks of the future
SG5 – Environment and climate change	**SG15** – Advanced networks
SG9 – Broad- and narrowcasting	**SG16** – Multimedia communcations
SG11 – Protocols and specifications of testing	**SG17** – Security issues

Fig. 2.5 Leading research groups ITU-T

Table 2.1 The main activities of Group 17

⬚ п/п	Activities of Group 17
Q1/17	Telecommunications/ICT security coordination
Q2/17	Security architecture
Q3/17	Information security management of telecommunications
Q4/17	Cybersecurity
Q5/17	Counteracts the spread of spam by technical means
Q6/17	Security aspects of general telecommunications services
Q7/17	Security of application services
Q8/17	Security of cloud computing
Q9/17	Telebiometrics
10/17	Identity of the management and device architecture
11/17	General technology to support application security
11/17	Formal languages for telecommunications software and testing

- Authentication and directory services (X.500, overview of conceptual models and services; X.509, basics of public key and certificates technology; etc.)
- System management (X.733, incident report function; X.740, security audit function; etc.)
- Basics of the security architecture (X.800, open systems infrastructure security architecture for ITU applications; X.802, security model of lower layers; X.803, security model of higher layers; etc.)
- Fax (T.36, security features when using fax machines of the third group; T.563, characteristics of terminals for use with facsimile machines of the fourth group; etc.)
- Television and cable systems (J.170, IPCabelcom security specifications; etc.)
- Security technology (X.841, information security objects for access control; etc.)
- Multimedia communications (H.233, confidentiality system for audiovisual services; H.234, management of encryption keys and authentication system in audiovisual services; etc.)

Table 2.2 gives a number of recommendations on cybersecurity.

Table 2.2 The recommendations of Group 17 on cybersecurity

SG 17 Question (2009–2013)	Recommendations			Harmonization phase
	Abbreviation	Full name	Harmonization equivalent	
4. Cybersecurity	X.1500 Appendix 1	Information on cybersecurity – Appendix 1. Methods for exchanging information on cybersecurity	No	Introduced
	X.1500 Amd.2	Information on cybersecurity – Amendment 2 - Corrections on cybersecurity information exchange methods	No	Introduced
	X.1500.1 (X. cybex.1)	Procedures for registering ARC facilities for the exchange of information on cybersecurity issues	No	Introduced
	X.1524 (X. cwe)	Numbering procedures	No	Introduced
	X.1526 (X. oval)	Vulnerability description Language	No	On Approval
4. Cybersecurity	X.1528 (X. cpe)	Numbering concept	No	Introduced
	X.1528.1 (X. cpe.1)	Assignable names concept	NISTIR 7695	Introduced
	X.1528.3 (X. cpe.3)	Definition of the general dictionary	NISTIR 7297	Introduced
	X.1528.4 (X. cpe.4)	Universal numbering language	NISTIR 7698	Introduced
	X.1541 (X. iodef)	Incident description formats	IETF RFC 5070	Introduced
	X.1544 (X. capec)	Classification of numbering attacks		On approval
	X.1580 (X. rid)	Network protection in real time	IETF RFC 6545	Introduced
	X.1581 (X. ridt)	Secure online exchanging in real time	IETF RFC 6546	Introduced

2.1.3 Recommendations: ITU-T X.800-X.849 Series

On September 15, 2012, ITU (*ITU-T*) adopted an international recommendation (Russia was involved in drafting the document), X. NCNS-1 ITU-T X.800-X.849 series – *Supplement on guidance for creating national IP-based public network security center for developing countries* – under the principle of "Cooperation for security." Implementing this recommendation makes it possible to design and implement a "single trust space" with managed collective security services

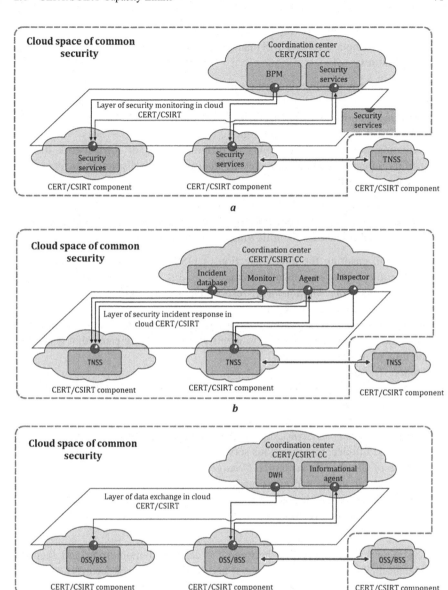

Fig. 2.6 Possible representations of the cloud space of common security

(including cloud services) on a voluntary basis. The main components of this trust space (control planes), namely, the service exchange plane, the security control plane, and the information exchange plane, are shown in Fig. 2.6a–c, respectively.

The recommendation notes that the telecommunications network includes either a single network management system (NMS) reflecting the status of the communication network or one or several *element management system* (*EMS*). Unlike EMS systems, a single NMS management system would permit the following security features:

- Monitoring communications between devices or communication lines in the network
- Reporting security incidents occurring at different nodes of the network, including different hardware manufacturers
- Predicting how security incidents at different nodes of the network disrupt stability, integrity, and security, causing malfunctions or disrupting the operation of whole network

NMS exchanges data with *Operation and Business Support Systems* (*OSS/BSS*) via *a server interface implemented using SNMP, OSS\J, MTOSI, COBRA*, etc.

Typically, OSS includes the following main components:

- *Mediation*, ensuring the interconnection of OSS/BSS solutions with heterogeneous equipment of various manufacturers
- *Resource/Inventory Management*, responsible for accounting of physical and logical resources of the network
- *Fault Management* as an alarm monitoring and management system, designed to filter and correlate them in order to identify the underlying cause that generated the flow of interrelated alarms
- *Trouble Ticketing Control*
- *SLA Management*, providing operational service monitoring available for internal and external users
- *Order Management*, required for tracking all stages of execution of the service order
- *Fraud Management Systems*, designed to interrupt and prevent unauthorized and unpaid use of telecommunications services
- *Module of Service Provisioning Management*, predicting the development of events and simulating various scenarios
- *Security Management*, providing control over access to network resources
- *Module of Accounting Management*, registering the time of use of various network resources

The algorithm of interaction between the centers of monitoring and security coordination (*Security Situation Centers*) and *NMS* (in terms of security) of significant operators implies the use of existing CYBEX formats, as well as their extensions. The adopted recommendation details the possible interaction protocols (languages). However, the rules for interaction between the national center and existing centers for responding to computer incidents (CERT, CSIRT) are discussed in the following recommendations [1–4].

Using these recommendations enables the design and implementation of a prospective cloud center for responding to computer security incident CERT/CSIRT within the national program "Information Society (2011–2020)." Here, support for end-to-end processes of interaction with collective security system participants is key to the project's success.

For each network security system to function properly and to reflect emerging threats (attacks), it is necessary to organize the information exchange between networks in the form of statistical data requests accumulated in each network's NMSVEMS (*OSS/BSS*). Such requests can also be received from the CERT/CSIRT based on the processing of the incidents database (usually the data is downloaded in XML format). Additionally, it is necessary to expand the formats described in the X.1500 series, as well as protocols to describe new incidents.

2.2 Example of Building a SOPCA

This section contains a review of current work toward creating an enterprise state and corporate segments of monitoring in the detection, prevention, and response to cyber-attacks (*SOPCA*) for the federal network of the Ministry of Education and Science of the Russian Federation. A promising approach to the creation of such a cyber-attack detection system based on domestic ViPNet technology has been proposed and justified [3, 5, 6].

2.2.1 Introduction

The relevance of designing a corporate segment of SOPCA in keeping with the aforementioned approaches to detecting, preventing, and eliminating the consequences of cyber-attacks owes to a simple reality: information security measures must be improved in light of the growing threats to information security [7–11].

The proposed system is designed for solving the following main tasks:

1. Detecting, preventing, and eliminating cyber-attacks' consequences, aimed at controlled educational information resources
2. Evaluating the degree of protection for the said resources
3. Determining the causes of the attack-related incidents
4. Collecting and analyzing data on the state of information security in controlled educational information resources, etc.

Throughout the work process, the requirements of the following normative documents were taken into account:

- Federal Law No. 149-FZ of July 27, 2006 (version of July 6, 2016) "On Information, Information Technologies and Information Protection"

- Federal Law of the Russian Federation of July 27, 2006 No. 152-FZ "On Personal Data (current version, 2016)"
- Doctrine of Information Security of the Russian Federation (approved by Presidential Decree No. 646 of December 5, 2016)
- Decree of the President of the Russian Federation of January 15, 2013, No. 31c "On the establishment of a state system for detecting, preventing and eliminating the consequences of cyber-attacks on Russia's information resources"
- Concept of a state system for detecting, preventing, and eliminating the consequences of cyber-attacks on information resources of the Russian Federation (approved by the President of the Russian Federation on December 12, 2014, No. K 1274)
- Order FSTEC of Russia of February 18, 2013, No. 21 "On the approval of the composition and content of organizational and technical measures to ensure the safety of personal data when processing them in information systems of personal data"
- Order FSTEC of Russia of February 11, 2013, No. 17 "On Approving the Requirements for the Protection of Information that Is not a State Secret Contained in State Information Systems"
- Order of the FSB of Russia and FSTEC of Russia on August 31, 2010, No. 416/489 "On approval of the requirements for the protection of information contained in public information systems"

2.2.2 Problem Solutions

Such authorial models, methods, and algorithms were applied to accomplish the assigned tasks [1–3, 5, 6, 12–51]:

- Static and dynamic software verification
- Computer systems and networks' security assessment
- Cyber-attack detection and prevention
- Information security incident investigation

In particular, a new method of handling cybersecurity incidents was introduced (Fig. 2.7).

The given method differs significantly from known analogues in its application of the first domestic signatures of "InfoTechs" company, which reflect the specifics of the cyber-attacks' implementation in Russian Federation. In this sense, "signatures" are particular text rules for detecting events in data transfer traffic based on many key parameters (number and string variables, operations on data [equality, algebraic, and bit operations], subsets of network addresses, etc.). For example, the above mentioned rule might look as follows (see Inset 2.1):

> **Inset 2.1 An Example of a Cyber-attack Signature**
> alert tcp $HOME_NET any -> $EXTERNAL_NET any (msg:»AM redirect
> AnglerURI landing page»; flow:to_server,established; content:»GET»;
> http_method; content:»Accept|3A|»; http_header; content:»*/*»; distance:1;
> http_header; content:» Referer|3A|»; http_header; pcre:»/^\V[a–z]+V
> (viewtopic|viewforum).php\? [a–z]=[0–9]+&[a–z]+=[0–9]+$/U»; pcre:»/
> Referer: http:VV([a–z.]+)[V].+?Host: (?:(?!\1).)*Connection/s»; flowbits:set,
> AnglerURI; flowbits:noalert; reference:url, symantec.com/security_response/
> attacksignatures/detail.jsp?asid=26992; classtype:client-side-exploit;
> sid:5002016; rev:24)

In this way, it was possible to quickly detect and process various "suspicious"
network packages with an abnormal structure containing redundant service data and
packages ostensibly intending to exploit vulnerabilities and malware.

2.2.3   Proposed Solution

The following components were included in the proposed solution for a cyber-attack
detection system on the departmental communication network of the Ministry of
Science and Education of Russian Federation: ViPNet IDS sensors, ViPNet Data
Forwarder control module, and ViPNet TIAS subsystem. The ViPNet IDS sensors
are meant for network traffic capture and its dynamic analysis. ViPNet Data Forwarder monitoring module collects and processes signs of primary and secondary
cyber-attack. The ViPNet Threats Intelligence Analytics System (ViPNet TIAS)

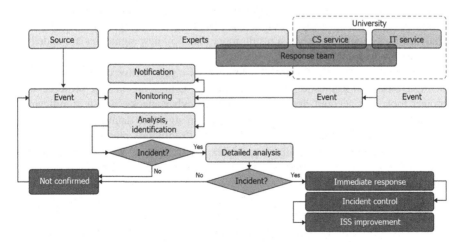

Fig. 2.7 Scheme of the analysis and algorithm for handling cybersecurity incidents

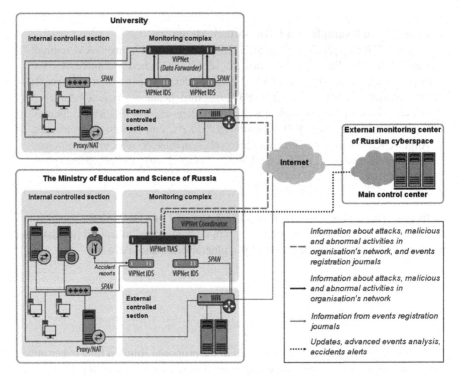

Fig. 2.8 The scheme of the proposed SOPCA solution

subsystem performs functions of cyber-attack detection, reporting on security incidents and their automated processing (Fig. 2.8).

In order to maintain the signature (Snort and AM rules) databases, which are responsible for reflecting the specifics of cyber-attacks in Russia, new malicious code samples and previously unknown types of group and massive cyber-attacks are analyzed on a regular basis. As a result, more than 150 new signatures are added to the database every month.

The results obtained led to the development of a prospective corporate segment SOPCA based on domestic VipNet technology. Unlike preexisting projects, this project took into account domestic specifics of cyber-attacks, and it significantly supplemented and verified the well-known foreign signature databases VRT (Cisco Systems, previously SourceFire) and ET Pro (Proofpoint, previously Emerging Threats). As a result, the ability of complex verification of decisive rules of the best foreign (more than 22,000 rules) and domestic practices (2500 rules) was first implemented.

In general, the project provides the ability to identify network attacks (intrusions) based on network traffic of the TCP/IP protocol stack analysis. To this end, the data analysis with a view to detecting intrusions is carried out by applying complex combinations of the best methods: signatory, correlation, invariant, and heuristic analysis. The project enables the following features of the SOPCA:

- Cyber-attack (intrusion) detection based on dynamic analysis of network traffic of the TCP/IP protocol stack for the protocols of all levels of the open systems interaction model (from the network to the applied)
- Cyber-attack registration (intrusions) in real (quasi-real) time
- Display of generalized statistical information about attacks
- Journaling of detected events and attacks for further analysis
- Selective event (attack) search according to the preset filters (time range, IP address, port, criticality degree)
- Attack (intrusion) journal export to a CSV file for further analysis in exterior applications
- Automated update mode without decisive rules
- Hardware-software complex ViPNet IDS masking as part of a supervised network
- Selective usage of certain detection rules or groups of rules at the discretion of security administrator
- Addition of new customized rules for the network traffic analysis
- Selective monitoring of network resources at the individual objects level
- Registration, display, and export of IP-packages that correspond to the registered events (attacks), to a PCAP file for further analysis in exterior software
- Automated SNMP transmission of generalized information about network attacks (intrusions) to a centralized monitoring system
- Executable and configuration files integrity control
- Downloadable attack detection rules bases integrity control

Most notably, the proposed project SOPCA allows analyzing the data transfer traffic on the 3–7 levels of OSI model. It also introduced a special preprocessor with a non-extendible set of rules aimed to detect attacks on the channel level. Additionally, it also implemented the ability to inspect HTTP, DCE/RPC, FTP, SMB, SMTP, and POP3 application protocol traffic. The ability to detect signs of cyber-attacks in the "raw" network packages in the event of modifications to network protocols or previously unknown protocols was implemented.

Panel 2.1 gives a sample technical report on the security status of the network segment of the Ministry of Science and Education of Russian Federation.

Panel 2.1 Sample Technical Report on the Security Status of the Network Segment of the Ministry of Science and Education of Russian Federation
In the first quarter of 2017 in a controlled network segment of the Ministry of Science and Education of Russian Federation (approximately 55,000 typical automated workplaces), out of 137,873,416 IS (information security) events, 98 significant CS (computer security) incidents were detected. In this usage,

(continued)

Panel 2.1 (continued)

"IS event" refers to the identified appearance of a certain system, network, or service state indicating a possible information security policy violation (or protective measures failure or a previously unknown safety-related situation). A "computer security incident" is the appearance of one (or several) undesirable information security event, leading to a cyber risk, for instance, the main activity operations compromised. In this, the sources of occurring security event messages appeared to be typical network devices and standard means of perimeter defense: network and host IDS, firewalls and routers, various network devices and sensors, security scanners, antivirus solutions, fraudulent systems or traps (honeypot), etc.

A significant increase, compared to the previous accounting period (fourth quarter of 2016), in attempts to scan and search for passwords to various information systems of organizations was recorded. Besides, the activity of malware has significantly increased (http://amonitoring.ru/service/security-operation-center/mssp/) (Fig. 2.9).

Event types description:

"Information events" refers to events with information that could prove useful for later event analysis.

"IS policy violations" refers to actions that violate the requirements of a supervised organization's IS policy.

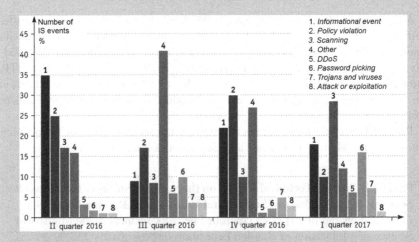

Fig. 2.9 Dynamics of information security events for the reporting period

(continued)

Panel 2.1 (continued)

An "attack or exploitation" refers to remote attempts at code execution or
   vulnerability exploitation on controlled resources.

"Scanning" refers to events marking network exploration before an attack
   attempt.

"Password picking" refers to attempts to access the controlled resources by
   authentication data selection.

"Trojans and viruses" are events, marking the infection of controlled resources
   by viruses or malware activity.

"DDoS" refers to attempts at distributed attacks on service denial.

"Other" refers to events not falling under any of the previously listed events.

According to cybersecurity event classification, the incidents identified had
the following distribution (Table 2.3).

Cybersecurity events distribution, according to days of the week, as shown
on Fig. 2.10

Table 2.3 CS incident classification distribution

Incident Class	Criticality			Incidents Total	Proportion of incidents
	High	Average	Low		
Malicious software	16	20	15	51	52%
Attack	9	5	1	15	16%
Passwords picking	11	3		14	14%
Violation of IS policy	2	4	3	9	9%
Exploitation of vulnerabilities	3	3		6	6%
DDoS	3			3	3%
Total				98	100%

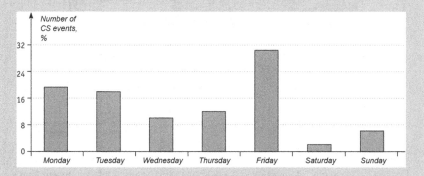

Fig. 2.10 Cybersecurity events distribution according to days of the week

(continued)

Panel 2.1 (continued)

In Fig. 2.11, a rise in the number of cybersecurity events is seen during the reporting period.

Cyber-attack sources

The location of the first hundred IP addresses, which were used to attack a controlled segment of the organization's network, is reflected in Fig. 2.12. The majority of the IP addresses shown are located in Russia, the USA, and Germany but this does not necessarily imply that the attackers were from these countries.

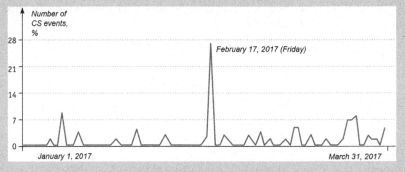

Fig. 2.11 Peak growth of cybersecurity incidents

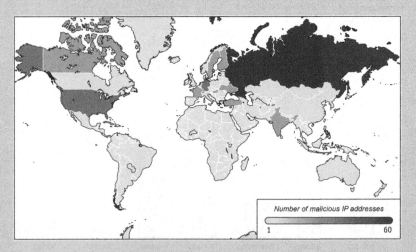

Fig. 2.12 Geography of cyber-attacks on a controlled segment of the organization's network

(continued)

Panel 2.1 (continued)
Cyber-attack targets and information-technical impact techniques

More than half of the incidents are attacks on users (Fig. 2.13).

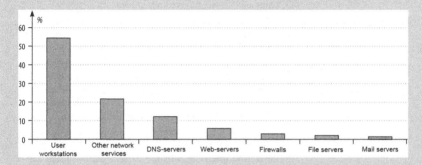

Fig. 2.13 Typical targets of cyber-attacks

Let us regard the techniques of information-technical impact used by the intruders (Table 2.4).

Sample detailed analysis of an information security event
Description of the performed services

Before beginning work on detection and prevention of cyber-attacks, those responsible for information security were asked to provide information about the protection of network segment of the Ministry of Science and Education of Russian Federation in order to configure the settings of the VipNet IDS hardware-software complex. As the required information was not received, event sources (hardware-software complex ViPNet IDS) were initiated and set to run with default settings. Additionally, the hardware-software complex was regularly updated with new signature databases.

Over the course of execution (more than 1.5 months), 2,700,000 events were analyzed (any system, service, or network phenomena identified suggesting possible IS policy violation or protective measures failure or an appearance of an unknown security situation), and 125 types of suspicious events and more than 50 types of IS critical events were detected (an appearance of one or more unwanted or unexpected IS events, which are associated with a high probability of activity discrediting and creating IS thread).

(continued)

Panel 2.1 (continued)

Table 2.4 Applied techniques of information-technical impact

Threat	Impact techniques
Advertising software	Terminal system infection, user information transfer to a command server, targeted ads shown
Password picking	Authentication data brute forcing for controlled organization's services and resource accessing – RDP, SSH, SMB, DB, web
IS policy violation	IS policy requirements violation (by users/administrators) in terms of old version or untrusted software. That software can be used by intruders to attack via vulnerability exploitation. Also, using company resources for their benefits (bitcoin/ethereum mining). Torrent-tracker usage
Viral software	Infection of a terminal system, virus spread on a local network, disabling/locking the services that prevent the virus from spreading, attempts to other attacks inside the network for critical data acquisition, and transfer to command servers
DDoS using the organization's resources	DDoS amplification – the technique of substituting one's own address with the victim's and generating small requests to open services. The service returns an answer to a request that is tens of times larger to a "sender's" address. An intruder performs a DDoS attack on a victim using a large number of resources from different organizations
Attempts to exploit vulnerabilities	The usage of deficiencies in a system for integrity violation and system malfunction. Vulnerability can be a programming error result, deficiencies in the design of a system, insecure passwords, viruses and other malware, script, and SQL injections. Some vulnerabilities are known only theoretically, while others are actively used and have known exploits
Web-resource page substitution	A hacker attack when the pages and important information are substituted by others, usually evocative (ads, warnings, threats, propaganda). Often, the access to the rest of the site is blocked or its previous content deleted

(continued)

Panel 2.1 (continued)
Event analysis

Over the course of the cyber-attack detection and warning, among all the analyzed events more than 100 of them were connected with malfunctioning network resources of the given controlled network segment as well as more than 40 events indicative of an IS policy violation (Table 2.5).

Table 2.5 Sample selection of suspicious events

Signature name	Description	Number
ET POLICY DNS Update From External net	DNS Update From External net	685 019
ET POLICY Vulnerable Java Version 7 and 8< update 66	Using an old version of Java	2913
ET WEB_SERVER Possible SQL Injection Attempt DELETE FROM	False alarm or invalid DB requests	922
ET WEB_SERVER SELECT USER SQL Injection Attempt in URI		922
ET POLICY Http Client Body contains passwd= in cleartext	Open text critical data transfer	5586
AM DNS Query for Suspicious amigobin.cdnmail.ru Domain	Sending requests to the site amigobin.cdnmail.ru. The site is considered potentially harmful	74
AM NETBIOS SMB Unicode share access	Access from the external network to an internal network resource through a NETBIOS SMB protocol	60078
AM NETBIOS SMB Unicode share access (with $)		374 978
GPL NETBIOS SMB-DS IPC$ Unicode share access		176 615
GPL NETBIOS SMB-DS Session Setup NTMLSSP Unicode asn1 overflow attempt		205 258
GPL NETBIOS SMB IPC$ Unicode share access		198 345
AM SMB2 repeated logon failure	Unsuccessful attempts to access SMB	2294
AM SNMP UDP Amplification Attack	Potential attack of a type "denial of service" through the SNMP protocol or incorrect network equipment configuration	379 727
AM SNMP Attempts to SNMP UDP Amplification Attack		662 878

(continued)

Panel 2.1 (continued)

IS events were analyzed using a corresponding automated security event registration system. As a result, more than 50 types of IS critical events were registered (e.g., events related to malware detection) (Table 2.6).

Table 2.6 Sample malware detection

Name	Infected hosts	Status	Description
AM TROJAN Multiple connections – possible Backdoor.Win32. DarkKomet	10.92.0.49	Expected	Probable activity of Trojan software Backdoor.Win32. DarkKomet which sends multiple requests to ports: 1995, 200, 6969, 1812, 2525, 1604
	10.92.0.49		
	10.91.0.14		
	10.91.0.31		
	10.92.0.31		
	10.92.0.48		

Currently, for timely and high-quality cyber-attack detection and warning, signature which has been verified (in terms of completeness, adequacy, and consistency) and completed for national specifics (*Snort, Cisco Talos, Emerging Threats* (*Open and Pro*), *Idappcom, Wurldtech*, etc.) may be used. Also corresponding virus signature databases from antivirus software producers can be useful (*Kaspersky Lab, Dr.Web, Freespace, AVG Free*, etc.). Methodological assistance can be provided by the data from the open vulnerability database of FSTEC Russia as well as other foreign companies and associations: RNA Virus Database, CVE (*Common Vulnerabilities and Exposures*), SecurityLab by Positive Technologies, CNews Bugtrack, Rapid, OSVDB (*Open-Source Vulnerability Database*), Exploit Database NVD (*National Vulnerabilities Database*) (http://nvd.nist.gov), etc.

Furthermore, to improve the completeness and reliability of information about IS, it is recommended to use data from feeds or available information sources:

- On network threads and anomalies – IRC, DNS, and IP (https://www.threatgrid.com) that support programming languages like Go, Ruby, Java, .NET, Perl, PHP, PowerShell, Python, RESTful, WSDL, and SOAP in formats JSON, XML, CyBOX, STiX, CSV, Snort, Raw, etc.
- On mixed thread types (IP and DNS addresses, host names, e-mails, URL and URI, hashes and file paths, CVE recordings and CIDR rules) (https://www.alienvault.com/open-threat-exchange)
- On network threads and anomalies – Arbor Active Threat Level Analysis System (ATLAS) (http://atlas.arbor.net/)
- On producer vulnerabilities – for example, Microsoft's bulletins, Cisco recommendations such as IntelliShield Security Information Service (https://www.cisco.com/security), etc.
- On domains that spread malicious code in AdBlock and ISA format – DNS-BH (Black Holing) (http://www.malwaredomains.com) project, etc.

- On spam and IP address threads (spammers, fishers, dangerous proxies, intercepted hosts, domains from spam) – DNSBL and ROKSO registry (http://www.spamhaus.org), etc.

Sample analysis from information sources is possible here: IOC (*Abuse.ch, Blocklist.de, Clean MX, Emerging Threats, Forensic Artifacts, Malware IOC, Nothink, Shadowserver*), DNS (*ISC DNS DB, BFK edv-consulting*), VirusShare. com, Crowd Strike, Farsight Security, Flashpoint Partners, IOCmap, iSightPartners, Microsoft CTIP, Mirrorma.com, ReversingLabs, SenderBase.org, Threat Recon, Team Cymru, Webroot, ZeusTracker, etc. It is possible to aggregate IS threats by indicators of compromise and given information about the intruders: OpenIOC (http://openioc.org), host signs; CybOX (http://cybox.mitre.org); OpenIOC и CybOX (https://github.com/CybOXProject/openioc-to-cybox); and STIX (http://stix.mitre.org), a description of threats, incidents, and intruders.

Some assistance can be provided by the recommendations, IODEF (RFC 5070) (http://www.ietf.org/rfc/rfc5070.txt), RFC 5901 (http://www.ietf.org/rfc/rfc5901. txt), IODEF-SCI, and x-arf (http://www.x-arf.org/), as well as information exchange: TAXII (http://taxii.mitre.org) using STIX; European standards VEDEF (http://www. terena.org/activities/tf-csirt/vedef.html) and SecDEF ENISA; CAIF (http://www. caif.info), DAF (http://www.cert-verbund.de/projects/daf.html), RID (RFC 6545/ 6546), and MANTIS (https://github.com/siemens/django-mantis.git) – OpenIOC association, CybOX, IODEF, STIX и TAXII, RFC 5941 – about fraud countering; and MMDEF (http://standards.ieee.org/develop/indconn/icsg/mmdef.html), in terms of malicious code, metadata exchange, etc.

For additional IS event exploration, the following tools can be used:

- Network telemetry collection and processing: Elastic Search, Log Stash, Kibana, Splunk, Security Onion, Flowplotter, Wireshark – tshark, Network Miner, Snort, Suricata, BRO, Flowbat, tcpdump, StealthWatch, etc.
- Search for indicators of compromise (IoC): Yara, PowerShell, AutoRuns – Utility, Loki, Wireshark – tshark, etc.
- TPL protocol (thread reports "coloring") support which gives an opportunity to detect the range of information spread
- CIF REN-ISAC (http://collectiveintel.net/) support for collecting and analyzing aggregated data and generating rules for Snort, Iptables, etc.

2.3   A Sample Hardware and Software Complex for the Cybersecurity Immune Protection System

This chapter considers a new method of countering cyber-attacks, the "immune response method." This method provides the fundamental capability of modeling the behavior dynamics of enterprise infrastructure under various types of attacker impact as well as identifying and rebuffing previously unknown attacks. The ability to make the security system adaptive and self-organizing offers another significant advantage, as it permits a flexible and appropriate response to cyber-attacks in real time. This chapter considers the basic ideas of the immune response method as well as its implementation in the proper counteraction system to cyber-attacks [14, 25, 33, 52–54].

2.3.1 Characteristics of the Research Direction

The following basic studies in the field of immune systems were forerunners in the method's development (Figs. 2.14 and 2.15).

- Hij and Cowell constructed an equation describing the change in the number of circulating antibodies as a function of the plasma cells number.
- Jilek proposed a probabilistic model range of the antigen interaction with an immunocompetent B cell and also simulated the Monte Carlo method of forming a clone originating from a single B cell.
- Bell, using the basic hypotheses of the clonal selection F. Bernet theory, constructed a mathematical model of the humoral immune response to a non-multiplying monovalent antigen. He also proposed a simple model of the immune response to the multiplying antigen, which describes the interaction between the antigen and the antibody.
- A qualitative study of the predator-prey model was carried out by Pimbley and then by Pimbley, Shu, and Kazarinov, after the introduction of the B-cell equation into the model. Similar model representations are being developed by Smirnova and Romanovsky.
- In 1974, Italian scientists Bruni, Jovenko, Koch, and Strem proposed a B-cell response model that describes the immunocyte population heterogeneity by means of continuous functions of two arguments: affinity and time. The main distinguishing model feature is the consideration of the immune response from

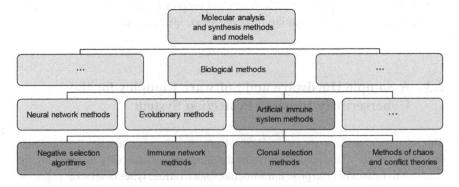

Fig. 2.14 Classification of methods of artificial immune systems

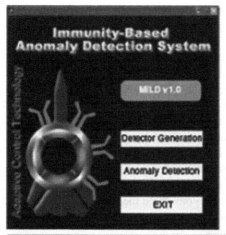

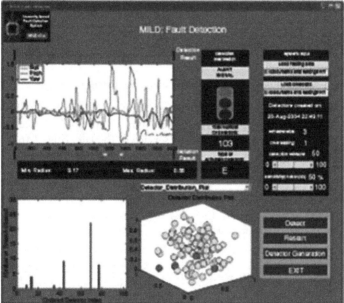

Fig. 2.15 Interface type of the immune response analysis system

the position of the bilinear system theory. The work was further developed in two directions. Moler modifies the model to describe a wider range of phenomena (production of antibodies of different classes, cooperation between T and B systems of immunity, etc.). On the other hand, these works are aimed at solving the problem of identifying the original model.

- G.I. Marchuk constructed and further specified the simplest mathematical model of an infectious disease, which is a system of ordinary nonlinear differential equations with retarded argument. In addition to the "antigen-antibody" reaction, the model describes the effect of antigen damage to the target organ on the immune process dynamics.

- Richter and Hoffman proposed original immune response models, based on Erne's network theory. The main attention in the models is given to the consideration of various events occurring in the network.
- Veltman and Butz described an immune response model using the threshold switching idea of a B lymphocyte from one state to another. The thresholds are introduced into the model equations as the delay times, which are the system state functions. Further model development was in the works of Gatic.
- Delisi examined the immune interaction mechanisms on the lymphocyte surface and suggested a tumor growth model in the body by analogy with the Bell model.
- Dibrov, Livshits, and Wolkenstein considered the simplest model of the humoral immune response, in which special attention was paid to the delay effect analysis on the immune process dynamics.
- Perlson considered the immune response from the position of the optimal control theory.
- Merrill proposed an immune response description from the catastrophe theory point.

The main practical developments in this area are listed below.

- The project Computational Immunology for Fraud Detection (CIFD)[14] was implemented in Great Britain. The project aims to develop a protection system based on *AIS* technology for the postal service of England (Fig. 2.16).

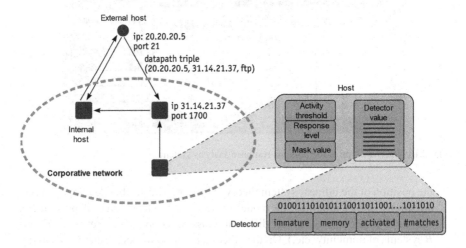

Fig. 2.16 Possible architecture of an artificial immune system

---

[14]http://www.icsa.ac.uk/CIFD

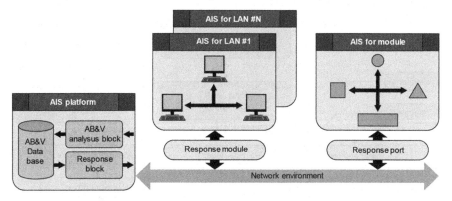

Fig. 2.17 Structural diagram of the immune system for cyber-attack detection

- In Europe, a project was implemented to develop a network-based system of detecting Lisys attacks by monitoring TCP SYN traffic. If suspicious TCP connections are detected, a notification is sent by e-mail. The Lisys consists of distributed 49-byte detectors that control the data path triple, such as the source and destination IP addresses, as well as the ports. At first, detectors are generated randomly, and those that correspond to normal traffic are gradually removed throughout the work process. In addition, the detectors have a lifetime, and as a result, the whole set of them, except those stored in memory, will be regenerated after a while. To reduce the false alarm number, an activation threshold is used, crossing which triggers the sensor to operate.
- In the USA, an extension to the Linux kernel – process homeostasis (pH), which allows detecting and, if necessary, slowing down the unusual behavior of application programs – was developed. To detect unusual behavior, at first, the system automatically creates call profiles made by different programs. It takes some time to create such a profile, after which the program can act independently, first including an exponential time delay for abnormal calls and then completely destroying the process. Since it is expensive and irrational to monitor all calls, the system works only with system calls that have full access to computer resources (Fig. 2.17).
- A similar project STIDE (*sequence time-delay embedding*) was designed to assist in detecting intrusions and recognizing unusual episodes of system calls. In the learning process, STIDE builds a database of all unique continuous system calls and then divides them into predetermined fixed-length parts. During operation, STIDE compares the obtained episodes having new traces with those already available in the database and reports an anomaly criterion indicating how many new calls differ from the norm (Fig. 2.18).

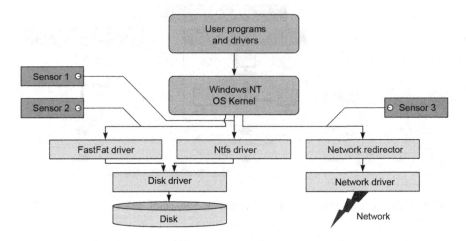

Use of sensors to determine the access correctness to the file system

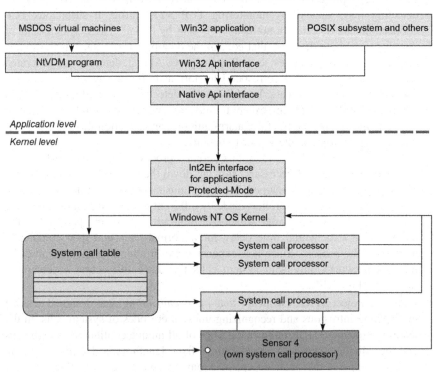

Use of call correctness sensor and application of the dynamically connected libraries

Fig. 2.18 The immune response technology elements

2.3.2   Mathematical Statement of the Problem

Initial Conditions

$$V(t^0) = V^0, C(t^0) = C^0, F(t^0) = F^0$$

Operating Conditions

$$\frac{dV}{dt} = (\beta - \gamma F)V,$$

where β is the coefficient of anomaly propagation in the system and γ is the number of detected anomalies at dt time.

$$\frac{dC}{dt} = \varepsilon(m)\alpha V(t - \tau)F(t - \tau) - \mu_C(C - C^*)$$

where $\varepsilon(m)$ is the characteristic of the *IS* functioning under the defeat of the main program subsystems, α is the coefficient characterizing the anomaly detection by the information security means, and μ_c is the coefficient that characterizes the lifetime of viruses before software updates.

Relative characteristic of the system damage:

$$\frac{dF}{dt} = \rho C - (\mu_f + \eta\gamma V)F,$$
$$\frac{dm}{dt} = \sigma V - \mu_m m,$$

where σV is the degree of damage to the *IS* and μ_m is the proportionality coefficient characterizing the value of the *IS* recovery period.

To Find

Immunological barrier, which characterizes the IS saturation by the means of countering computer attacks:

$$V^* = \frac{\mu_f(\gamma + F^*V)}{\beta\eta\gamma} > V^0 > 0$$

2.3.2.1   The Main Ideas of the Proposed Method

To solve the task, the following model of the malware impact on the *IS* operating environment was developed (Fig. 2.19).

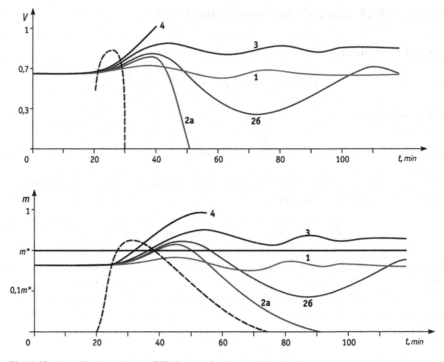

Fig. 2.19 A qualitative picture of IS damage by the combined malware

$$\frac{dV_i}{dt} = (\beta_i - \gamma_i F_i)V_i,$$

$$\frac{dF_i}{dt} = q_i C_i - \eta_i \gamma_i F_i V_i - \mu_i F_i,$$

$$\frac{dC_i}{dt} = \xi p_s(V_i)\alpha_i F_i(t - \tau) \times V_i(t - \tau) - \mu_{C_i}(C_i - C_i^*), \qquad (2.1)$$

$$\frac{dm_i}{dt} = \sigma_i V_i - \mu_{m_i} m_i$$

where

$i = \overline{1, N}$ is the "number" of the cyber-attack types.

N is the number of different cyber-attack types.

$V_i(t)$ is the concentration of *i-th* malware in the total volume of the required *IS* functions.

$F_i(t)$ is the concentration of antibodies specific to the *i-th* malicious software.

$C_i(t)$ is the concentration of detection countermeasures to the i-th cyber-attack.

$m_i(t)$ is the relative characteristic of the IS defeating by the i-th attack, $0 \leq m_i \leq 1$.

$\xi = \prod\limits_{i=1}^{N} \xi_i(m_i)$ is the function that characterizes the general IS state.

$\xi_i(m_i)$ is a nonincreasing continuous function characterizing the general IS state for the i-th attack, $\xi_i(0) = 1$, $\xi_i(1) = 0$, $0 \leq \xi_i(m_i) \leq 1$.

To the system of Eq. (2.1), we add the initial data for $t = t^0$.

$$V(t^0) = V^0, C(t^0) = C^0$$

1. The concentration of breeding malware antigens is $V(t)$.
2. The antibody concentration $F(t)$ (antibodies-substrates of the immune system of the IS operating environment, neutralizing antigens).
3. The measure concentration to monitor and to prevent the malicious software $C(t)$ effects.
4. Relative characteristic of the damage to the IS system environment $m(t)$.

As a result, the following main assertions were deduced.

Assertion 1 If nonnegative initial data for $t = t^0 = 0$

$$V^0 \geq 0, C^0 \geq 0, F^0 \geq 0, m^0 \geq 0$$

The solution of problem (1), (2) exists and is unique for all $t \geq 0$.

Assertion 2 For all $t \geq 0$, the solution of problem (1), (2) is continuous and nonnegative:

$$V(t) \geq 0, C(t) \geq 0, F(t) \geq 0, m(t) \geq 0.$$

Assertion 3 The existence and uniqueness theorem of problem (1), (2) solution allows obtaining formal models of IS damage by malware.

Assertion 4 The impact of malicious software, which does not lead to a stability loss of the IS functioning, satisfies the inequality:

$$0 < V_0 < \frac{\mu_f(\gamma F^* - b)}{\beta \eta \gamma} = V^*$$

Assertion 5 The V^* value is the immunological barrier of the IS operating environment. The immunological barrier is passed if the software impacts V^0 satisfy the condition $V^0 > V^*$ and is not passed otherwise (Fig. 2.20).

Fig. 2.20 Area of
admissible solutions

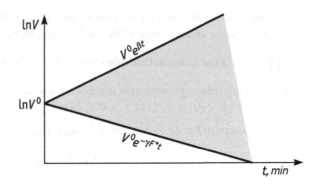

Fig. 2.20 Area of
admissible solutions

2.3.3   The Main Algorithms of the Immune Response Method

Figure 2.21 shows the problems whose solution involved practical implementation
of the given method.

As a result, five algorithms were developed to solve problems of the impact
analysis, detection, rebuffing, learning, and evaluation. Below, there are two of
them: detection and learning.

2.3.4   Detection Algorithm

The detection algorithm is schematically shown in Fig. 2.22.

Given

1. Observed multidimensional trajectory X (t), containing data received from the
   system sensors.
2. Set of B classes of the system abnormal behavior. For each class $b \in B$, a
   reference trajectory Y_{Anom}^b is given, and the trajectories of the different classes
   of abnormal behavior do not intersect.
3. Limitations on the recognition completeness and accuracy in the form of restrictions on the number of recognition and classification errors:

$$el \leq val1 \text{ and } e2 \leq val2$$

where el is the number of recognition errors of the first kind, $e2$ is the number of
recognition errors of the second kind, and $val1$ and $val2$ are the given numerical
constraints.

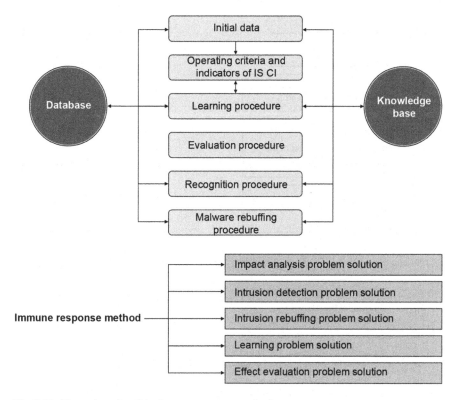

Fig. 2.21 The main tasks of the immune response method

Required

Considering the limitations on completeness and accuracy, it is necessary to recognize and classify the abnormal behavior in the system operation based on the observed trajectory $X(t)$ and the set of standard trajectories

$$\{Y_{Anom}\} = \bigcup_{b \in B} Y_{Anom}^{b}$$

Restrictions

The total number of recognition errors on the learning sample should not exceed the specified parameter

$$P_{wr} : v\left(A, \tilde{X}\right) \le P_{wr}$$

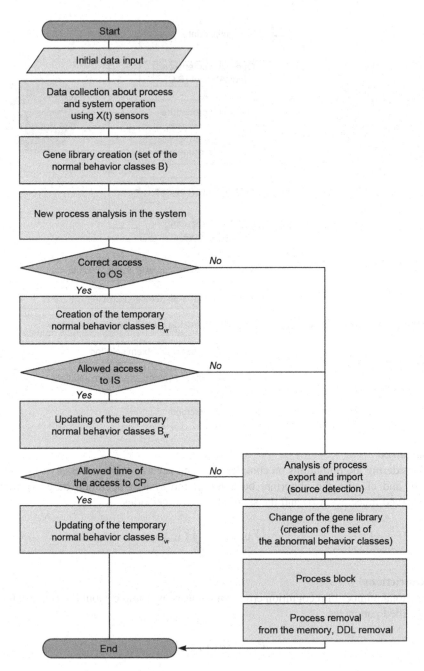

Fig. 2.22 Cyber-attack detection algorithm

Requirements to the Generalizing Ability

The algorithm A must be resilient with a given probability P_G to the possible distortion functions $\{G_1 \ldots, G_s\}$ of the sample trajectories if these distortions are given:

$$\forall_i, \forall_j \in [1, S] : P\left[A\left(X_{ir}^i\right) = A\left(G_j\left(X_{ir}^i\right)\right)\right] \geq P_G$$

where $X_{ir}^k = G_j\left(X_{ir}^i\right)$ is the trajectory deduced from X_{ir}^i after distortion by the G_j function.

Universal Restrictions

A algorithm must be able to generalize, i.e., to show good results not only on the learning $\tilde{X}$ selection but also on the entire set of X trajectories. To do this, it must minimize the target function $\psi(o1, o2)$ in the control $\tilde{X}$ selection, where $o1$ and $o2$ are the number of recognition errors of the first and second kinds.

Limitations on Computational Complexity

The computational complexity of the recognition algorithm $\Theta(A)$ must be limited to the predefined function $Ef\ (l,\ m)$, which is determined by the structure and characteristics of the calculator used: $\Theta(A) \leq Ef\ (l,\ m)$, where $Ef\ (l,\ m)$ is a function of l-number of sample trajectories and m-maximum length of the sample trajectory.

2.3.5   Learning Algorithm

We give a verbal description of the algorithm shown in Fig. 2.23:

1. From the original trajectory $X = (x_1 \ldots, x_n)$, we pass to the sequence of axioms $J = (j_1, j_2 \ldots, j_n)$, where j_i is the number of the comparable i axiom. The sample trajectories $\{Y_{Anom}\}$, corresponding to different classes of abnormal behavior, are marked.
2. In the marking range J, the sequences of axioms, corresponding to the markings of the sample trajectories, are sought.
3. Thus, abnormality of the observed system is not determined as a result by searching the sample $\{Y_{Anom}\}$ trajectories in observed X trajectory but by searching the markings of the sample trajectories in the marking range J.

The stopping criterion of the learning algorithm within the framework of the template is the following integral criterions:

- The conditions of the recognition algorithm learning problem, listed above, are fulfilled (in this case the algorithm is considered to be successful).
- The total number of algorithm iterations exceeded the predetermined value in advance or the number of iterations exceeded the specified parameter without improving the solution (in this case, the algorithm is considered to be unsuccessful).

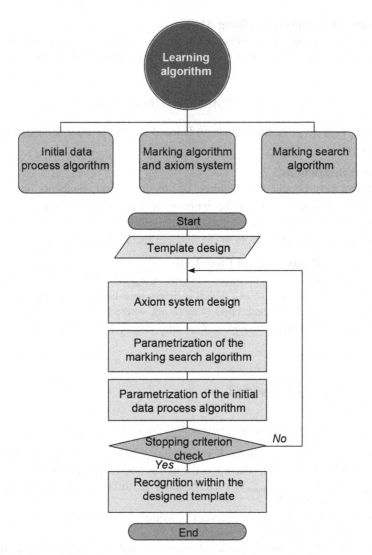

Fig. 2.23 Learning algorithm of the immune system for cyber-attack detection

Additional restrictions:

- The completeness condition, which means that for any point of an admissible trajectory there is an axiom from the system of axioms which marks it
- The uniqueness condition, which is in fact that any point of an admissible trajectory can be marked only by one axiom from the system of axioms

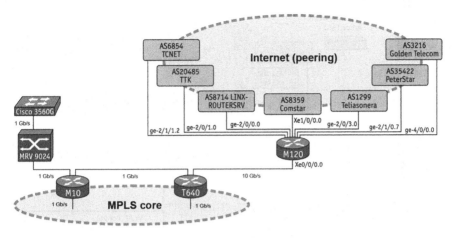

Fig. 2.24 Demonstration stand layout

2.3.6 Immune Response Method Implementation

To implement the immune response method and to prove the reliability of the obtained results, full-scale tests were carried out at the next enterprise site (Fig. 2.24).

Here, the network core is an MPLS ring formed by Juniper Networks M- and T-series routers. The main traffic is concentrated on the T640 and comes/returns by the 10 GB/s interface to peering partners from the border router M120. The diagram shows seven existing links with other autonomous systems. The main traffic comes from Comstar that is also connected by a 10 GB/s interface.

According to the statistics collected, the interface between the M120 and the T640 is approximately 10% at various moments in time; traffic observed on it ranges from 900 to 1100 MB/s.

The current load of the Routing Engine on the M120 router is less than a percent. Thus, it can be stated that with 10% interface utilization, the M120 has a solid performance margin for solving problems associated with the immune response method implementation.

The scheme of an artificial immune system to counteract cyber-attacks is shown in Fig. 2.25. Considering that the port cost on the M120 router is relatively high, it was decided to terminate the device Arbor Peakflow SP CP5 and Arbor Peakflow TMS interfaces, adapted for the immune response, on a Cisco 3560G commutator that has a relevant port capacity and connected via MRV to the M10 PE-router. The interfaces on all commutators are gigabit, and the connection between M10 and T640 is also gigabit. Load measurements of the connection showed that it was loaded approximately on 30 MB/s. Accordingly, there remains a sufficient reserve for the implementation of the immune response method, the service traffic analysis required for the operation of an artificial immune system to counteract computer attacks (flows, control traffic, attack traffic), and return of sanitized traffic to the

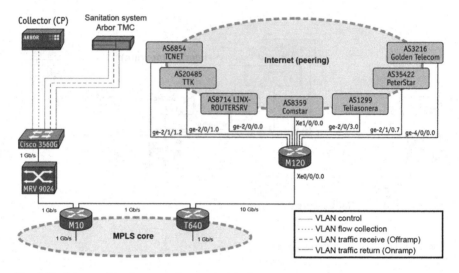

Fig. 2.25 Scheme of the demonstration stand of the immune response system

network. Calculation of the total extra costs for the service information transfer of the artificial immune system indicates that, even taking into account the amount of processed data (about 10 GB/s), the expenses should not exceed 500 Mb/s.

It should be noted that on each of the Arbor devices (CP and TMS), there are several physical ports. They are configured as L3 interfaces and can be used to manage and collect the flows from other routers, analyze traffic, and receive/return traffic when it is sanitized. It is proposed to organize a separate VLAN, which administrators will have access to manage the corresponding CP and TMS devices. Actually, the management console is "located" on the Arbor CP device, which is connected to the TMS device, so it is possible to create a centralized security administrator workstation.

2.3.7 General Operation Algorithm

The immune response system algorithm is reduced to the following basic steps:

- Assessing state security based on statistical data
- Forecasting the situation
- Detecting cyber-attacks
- Rebuffing cyber-attacks
- Preventing cyber-attacks

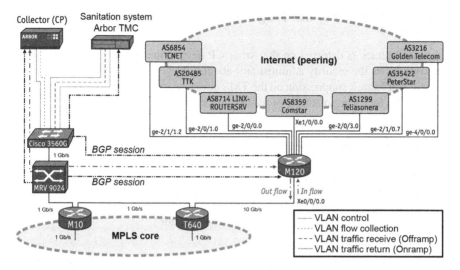

Fig. 2.26 Implementation of the immune response algorithm

For testing purposes within a pilot zone to analyze the traffic statistics and the effect on network flows, one border M120 router is available because it collects all channels and is used for communication with the network core. The remaining two routers participate in the work of MPLS ring and transmit the flows from M120 further through the network. However, these routers do not deal with the routing table as such. Therefore, a BGP session is established on the M120 router from both nodes of the Arbor Peakflow SP CP and Arbor TMS from control interfaces. The first uses them to analyze the BGP table state (its stability) and the second to generate updates for the router, if necessary, to redirect flows for sanitation (Fig. 2.26).

Since in order to obtain reliable and consistent statistics, the immune response system should see the maximum information amount (at least "symmetric" data exchange), it is proposed to activate the statistical data generation for the Core interface on the input and output. This will permit monitoring all the data that comes from the Internet and goes back. Generated flows are collected on the Arbor CP device and processed for deviation detection. If necessary, gene and pattern libraries of standard normal system behavior are connected.

Also, to get an interface list and counters from them on the Arbor CP, it is required to configure the M120 poll on SNMP.

The main task in the process of deploying the immune response system is to ensure the traffic routing correctness under network anomaly detection, redirecting the flows to the TMS and returning the sanitized traffic to the nearest router. This necessarily involved providing for elimination mechanisms of routing loops, which arise when redirecting traffic to and from sanitation.

2.3.8 Algorithm of the Traffic Filtering in Attack Mode

If a cyber-attack is detected on the Arbor CP device, the artificial immune system
sensor notifies the security administrator about this event and allows the sensors of
"intelligent" attack suppression on the TMS device to be activated. In this case, TMS
can use the following methods of traffic sanitation:

- Global exception list
- Per mitigation filters
- HTTP mitigation (HTTP request limiting, HTTP object limiting)
- Zombie removal
- TCP SYN authentication
- TCP connection reset
- DNS (malformed DNS filtering, DNS authentication)
- Baseline enforcement

In any case, incorrect traffic is identified by a predetermined length prefix, which
is advertised to peer routers to redirect the flow to the device off-ramp interface.

For example, in the proposed scheme (Fig. 2.27), when the response procedure is
started, the TMS generates a BGP update, where it advertises the attacked prefix via
its off-ramp interface. After that, the traffic coming to M120 from the Internet is
redirected to TMS, where one of the immune response methods listed above is used
to "sanitize" it.

After removing the attack traces from the incoming traffic, the TMS must return it
to the "backbone." For this, the on-ramp interface is used (off-ramp and on-ramp
interfaces can be physically combined). As an option for "return" traffic, the
following options can be suggested:

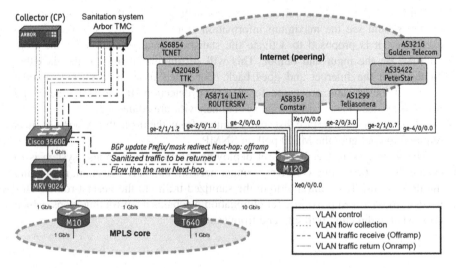

Fig. 2.27 Algorithm scheme for intelligent attack suppression

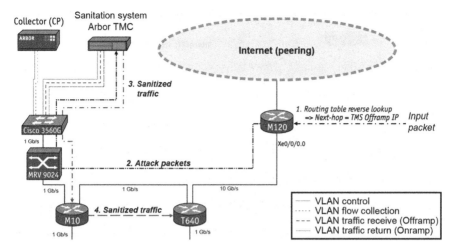

Fig. 2.28 Behavior scheme in the mode of attack detection

- Applications of physical on-ramp interface, different from the original off-ramp with other logical (IP) addressing
- Application of GRE tunnels to transfer traffic to the "the network exit" point, i.e., to the CE router.

The main task of any of these sanitation traffic methods is to return the latter in such a way that would exclude its entry to the original router that provided off-ramping traffic on the TMS; otherwise a routing loop will inevitably arise. It usually suffices to allocate different physical network routers.

To return traffic to the existing immune system scheme, two options are possible. The first is the principle possibility of returning traffic to one of the routers working in the MPLS core network, meaning M10.

The process of returning packets in such cases is depicted in Fig. 2.28.

When an attack has been detected and a command sent from the CP to initiate sanitation, TMS sends a BGP update to the M120 with the known community tag "NO-ADVERTISE." Accordingly, the route is modified only on the M120, instructing it to forward the information to the off-ramp TMS interface. After the traffic sanitation, the latter returns traffic to the on-ramp interface, connected via the VLAN with the interface to the M10.

Since the table on M10 is unmodified, the packet goes further to the subscriber.

The second variant is alternative and works if there is only one router (M120), and the guarantee routing without loops should be reached by other methods. For example, policy-based traffic routing could be implemented as follows. The traffic comes with on-ramp VLAN so that the M120 does not do the next-hop reverse lookup on the routing table but immediately sends the packet to the output interface toward the T640.

In this case, the package route will be as shown in Fig. 2.29.

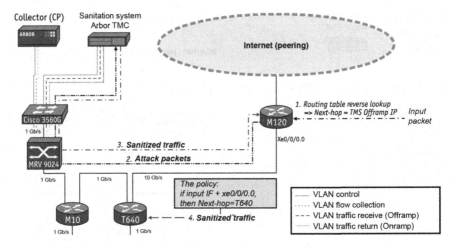

Fig. 2.29 Routing scheme for traffic return

A unique characteristic of the proposed solution is its implementation of the routing mechanism based on appropriate policies. In this way, it ensures the packet's passing without loops in the event of choosing an alternative version of their return via on-ramp interface. However, even if the artificial immune system, based on Arbor solutions, fails, this will not affect the core network performance. The only major change in the packet path is the policy on the interface (VLAN) connected to the on-ramp-based TMS interface, which is used only by TMS. For this reason, there should not be any other "random" packets.

As already noted, the current load of the M120 router processor does not exceed 1% on average, so assigning it the tasks to generate routing engine flows will not lead to a significant slowdown. The main task is to choose the sampling ratio correctly. It is assumed that this will not be more than 1/1000 (which is the value recommended by Arbor Networks). To err on the side of safety, you can start with a ratio of 1/10000.

As for the communication channel bandwidth, the main load will be on the channel between the T640 and M10, which is used at about 5–10% of total capacity. In this case through the interfaces between the router and the immune system, based on the Arbor solutions, the following data is transmitted:

- SNMP polling
- Flow statistics
- BGP view (in the first stage)
- Traffic redirected for sanitation (only at the time of the attack and only on the prefixes that are set)
- Service information (gene library, counteraction procedures, etc.)

At the first stage of method implementation, the Arbor CP system was installed, set up, and adapted for implementing the immune response method. Then, after accumulating the necessary statistics, the immune suppression system of TMS attacks was used to detect computer attacks.

2.3.9 The Immune System Work Example

ARBOR PeakFlow SP data collection devices, adapted to the implementation of the immune response method, detect anomalies in the network and redirect the received information to the ARBOR PeakFlow SP CP controller. The latter, in turn, analyzes the information, uses the gene and antibody library, and, if necessary, automatically/ manually activates appropriate information protection means (e.g., ARBOR PeakFlow SP TMS). The tool checks and filters the packets. With a positive filtering result, the filtered traffic is re-injected back into the network (Fig. 2.30).

After successful attack blocking, a "pending" analysis phase occurs, where the statistics collection tools and the event history analysis can be used. The examples of the tools are Arbor Peakflow, Juniper IDP (*Netscreen Security Manager*), and, if necessary, specialized postmortem analysis based on Gigon and Network General solutions, as well as ArcSight solutions.

2.3.10 Effectiveness Evaluation

The effectiveness was evaluated in two different ways:

- Evaluation of the intellectual system effect of counteracting computer attacks based on the immune response method implementation (Fig. 2.31)
- Evaluation of the immune response system effect on the main quality indicators and the functional stability of the enterprise IS (see Table 2.7)

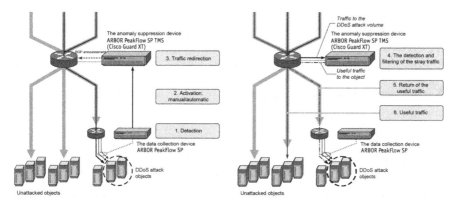

Fig. 2.30 Traffic analysis scheme

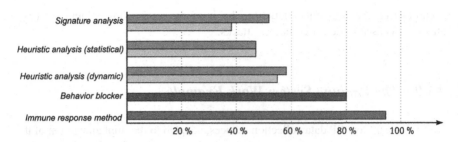

Fig. 2.31 Evaluation of the effect of counteracting cyber-attacks based on the immune response method implementation

Accuracy of detection:

$$\alpha = \phi^1_{\sum} / N^1_a$$

where:

α is the detection accuracy index.
N^1_a is the total number of analysis operations.
$\phi^1_{\sum}$ is the total number of errors, $\phi^1_{\sum} = \phi^2_f + \phi^3_m$.
ϕ^2_f is the number of false responses.
ϕ^3_m is the number of missed attacks.

$$\alpha^1_{\sum} = \sum_{i=1}^{K} \alpha^2_i \frac{N^2_i}{N^3_{\sum}}$$

where:

$\alpha^1_{\sum}$ is the generalized detection accuracy index.
α^2_i is an analysis component.

$N^3_{\sum}$ is the total number of analysis operations in $N^3_{\sum} = \sum_{i=1}^{K} N^2_i$.

N^2_i is the number of analysis operations.
K is the number of analysis components.

Table 2.7 Evaluation of the immune system influence against the attacks on the IS

Performance characteristics of IS operation	The quality and stability indicators of the IS operation	Valid values that characterize the level of system integrity	
		At an increased risk	At an acceptable risk
Reliability of the information submission, requested or issued forcefully (execution of specified technological operations)	Average operating time of the object to deny or to fail (T_{op})	–	–
	The average recovery time for an object after a denial or failure (T_{rec})	–	–
	The object availability coefficient (K_a)	0.999	0.9995
	Probability of the reliable reporting and/or communication of the requested (issued forcefully) output data (P_{inf}) for a given period of the IS operation (T_g)	0.99	0.99
	The probability of reliable performance of technological operations (P_{rel}) for a given period of operation of IP (T_g)	0.99	0.99
Timeliness of information submission, requested or issued compulsorily (execution of specified technological operations)	The average system response time during request processing and/or information delivery (T_{full}) or the probability of processing information (P_{tim}) in the given time (T_g)	0.90	0.95
	The average time to complete a process operation (T_{full}) or the probability of performing a process operation (P_{tim}) in the given time (T_g)	0.89	0.91
Completeness of used information	The probability of ensuring the completeness of the operative introduction to the IS the new real-world objects of considering the domain (P_{full})	0.8 – CRV 0.7	0.9 – CRV 0.8
Relevance of information used	The probability of the continued relevance of the information at the time of its use, (P_{rel})	0.95	0.99
The data accuracy after control	The probability ($P_{er\ ab}$) of the error absence in the input data on paper medium with permissible time for the control procedure (T_g)	0.95	0.97
	Probability (P_{com}) of the error absence in the input data on the computer medium with	0.97	0.99

(continued)

Table 2.7 (continued)

Performance characteristics of IS operation	The quality and stability indicators of the IS operation	Valid values that characterize the level of system integrity	
		At an increased risk	At an acceptable risk
	permissible time for the control procedure (T_g)		
Correctness of information processing	The probability (P_{corr}) of obtaining the correct results of information processing in the given time (T_g)	0.99	0.99
Officials actions accuracy	The probability of error-free actions of the officials (P_{people}) in the given operation period (T_g)	0.90	0.95
Security against dangerous software and hardware impacts	The probability of the absence of a hazardous effect (P_{eff}) in the given operation period (T_g)	0.99	0.99

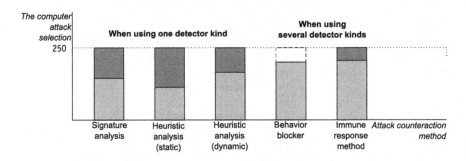

By detection completeness:

$$\varpi = A_{det}/A_{\sum}$$

where:

ϖ is the index of detection completeness.
A_{det} is the number of attacks detected.
$A_{\sum}$ is the total number of cyber-attacks.

The results of full-scale experiments, as well as simulation modeling of the immune response method, revealed the following dependencies of the method and algorithms of the immune response implementation (Figs. 2.32, 2.33, 2.34, and 2.35).

Fig. 2.32 Dependence of the dynamics of the concentration of malware V(t) on the dose of IS damage for $\beta = \gamma F^*$ in the case of (**a**) a "normal immune system" and (**b**) "immunodeficiency." Here V^* is the immunological barrier value of the IS operating environment

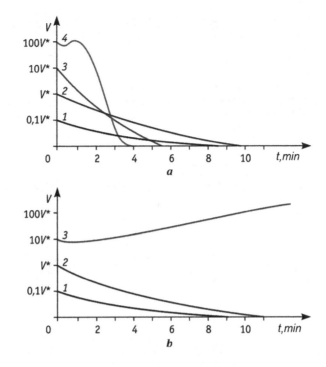

Fig. 2.33 (a) Dynamics of malware concentration in case of IS damage depending on the damage intensity $\beta(\beta_1 > \beta_2 > \beta_3 > \beta_4)$. (b) The phases and significance of the IS damage by malware in changing the IS damage coefficient $\sigma(\sigma_1 > \sigma_2 > \sigma_3 > \sigma_4)$. (c) Dynamics of malware concentration from $V^0(V_1^0 > V_2^0 > V_3^0 > V_4^0)$

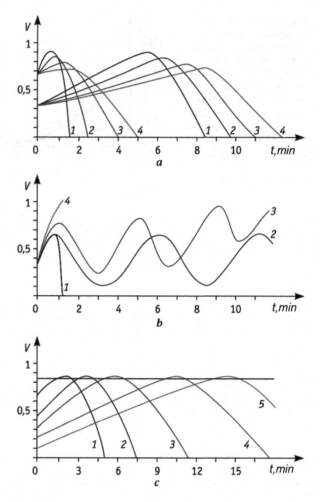

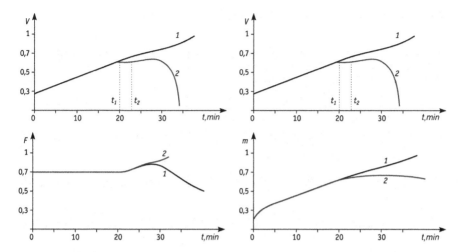

Fig. 2.34 Modeling the malicious software rebuffing by detecting and neutralizing rootkits in the interval $t_1 \leq t \leq t_2$

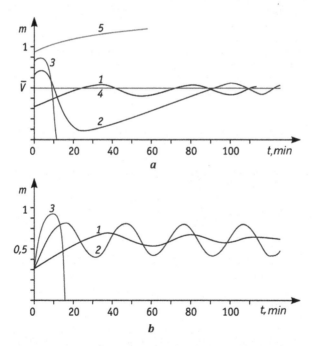

Fig. 2.35 The dynamics of the concentration of malicious software in the IS, depending on (**a**) *the dose of damage* $V^0 (V_1^0 > V_2^0 > V_3^0 > V_4^0 = \bar{V}(V_5^0))$, (**b**) *the rate of multiplication of malicious software* $\beta(\beta_1 > \beta_2 > \beta_3)$

References

1. Petrenko, A.S., Petrenko, S.A.: Super-productive monitoring centers for security threats. Part 1. Protect. Inf. Inside. **2**(74), 29–36 (2017)
2. Petrenko, A.S., Bugaev, I.A., Petrenko, S.A.: Master data management system SOPKA. Inf. Protect. Inside. **5**(71), 37–43 (2016)
3. Petrenko, A.S., Petrenko, S.A.: Large data technologies (BigData) in the field of information security. Inf. Protect. Inside. **4**(70), 82–88 (2016)
4. Portnoy, L., et al.: Intrusion detection with unlabeled data using clustering. ACM Workshop on Data Mining Applied to Security (2001)
5. Petrenko, A.S., Petrenko, S.A.: Designing of corporate segment SOPKA. Protect. Inf. Inside. **6** (72), 48–50 (2016)
6. Petrenko, S.A., Asadullin, A.Y., Petrenko, A.S.: Evolution of the von Neumann architecture. Protect. Inf. Inside. **2**(74), 18–28 (2017)
7. Abramov, S.M.: History of development and implementation of a series of Russian supercomputers with cluster architecture. In: History of Domestic Electronic Computers. 2nd edn, Rev. and additional; color. Ill.: Publishing house "Capital Encyclopedia", Moscow (2016)
8. Active Engagement, Modern Defence. Strategic Concept for the Defence and Security of the Members of the North Atlantic Treaty Organisation adopted by Heads of State and Government in Lisbon. November 19, 2010 [Electronic resource]. Access mode: http://www.nato.int/cps/en/SID-14EF0623-198FC77E/natolive/official_texts_68580.htm
9. Advances in the field of information and telecommunications in the context of international security. Report of the First Committee. Document A/66/407 dated November 10, 2011 [Electronic resource]. Access mode: http://www.un.org/en/documents/ods.asp?m=A/66/407
10. Agreement between the governments of the member states of the Shanghai Cooperation Organization on cooperation in the field of international information security from June 16, 2009, Yekaterinburg. Appendix 1. [Electronic resource]. Access mode: https://ccdcoe.org/sites/default/files/documents/SCO-090616-IISAgreementRussian.pdf
11. Almgren, M.: Consolidation and evaluation of IDS taxonomies. In: Proceedings of the Eight Nordic Workshop on Secure IT Systems, NordSec 2003
12. Barabanov, A.V., Markov, A.S., Tsirlov, V.L.: Methodological framework for analysis and synthesis of a set of secure software development controls. J. Theor. Appl. Info. Technol. **88**(1), 77–88 (2016)
13. Barabanov, A., Lavrov, A., Markov, A., Polotnyanschikov, I., Tsirlov, V.: The study into cross-site request forgery attacks within the framework of analysis of software vulnerabilities. In: Preliminary proceedings of the 11th Spring/Summer Young Researchers' Colloquium on Software Engineering (Innopolis, Republic of Tatarstan, Russian Federation, June 5–7, 2017), pp. 105–109. SYRCoSE, ISP RAS
14. Biryukov, D.N.: Cognitive-functional memory specification for simulation of purposeful behavior of cyber systems. Proc. SPIIRAS. **3**(40), 55–76 (2015)
15. Biryukov, D.N., Lomako, A.G.: Denotational semantics of knowledge contexts in ontological modeling of the subject areas of conflict. Proc. SPIIRAS. **5**(42), 155–179 (2015)
16. Biryukov, D.N., Glukhov, A.P., Pilkevich, S.V., Sabirov, T.R.: Approach to the processing of knowledge in the memory of an intellectual system. Natur. Tech. Sci. **11**, 455–466 (2015)
17. Biryukov, D.N., Lomako, A.G.: Approach to the construction of information security systems capable of synthesizing scenarios of anticipatory behavior in the information conflict. Protect. Inf. Inside. **6**(60), 42–50 (2014)
18. Biryukov, D.N., Lomako, A.G.: The formalization of semantics for representation of knowledge about the behavior of conflicting parties: materials of the 22nd scientific-practical conference "Methods and technical means of information security", pp. 8–11. Publishing house of Polytechnic University, St. Petersburg (2013)

19. Biryukov, D.N., Lomako, A.G., Petrenko, S.A.: Generating scenarios for preventing cyber-attacks. Protect. Inf. Inside. **4**(76) (2017)
20. Biryukov, D.N., Lomako, A.G., Rostovtsev, Y.G.: The appearance of anticipatory systems to prevent the risks of cyber threat realization. Proc. SPIIRAS. **2**(39), 5–25 (2015)
21. Biryukov, D.N., Lomako, A.G., Sabirov, T.R.: Multilevel Modeling of Pre-Emptive Behavior Scenarios. Problems of Information Security. Computer systems, vol. 4, pp. 41–50. Publishing house of Polytechnic University, St. Petersburg (2014)
22. Biryukov, D.N., Rostovtsev, Y.G.: Approach to constructing a consistent theory of synthesis of scenarios of anticipatory behavior in a conflict. Proc. SPIIRAS. **1**(38), 94–111 (2015)
23. Biryukov, D.N., Lomako, A.G.: Approach to Building a Cyber Threat Prevention System. Problems of Information Security. Computer systems, vol. 2, pp. 13–19. Publishing house of Polytechnic University, St. Petersburg (2013)
24. Bocharov, V.A., Markin, V.I.: Fundamentals of Logic. Moscow State University, Moscow (2008)
25. Kotenko, I.V.: Intellectual mechanisms of cybersecurity management. Proceedings of ISA RAS. Risk Manag. Safety, **41**, 74–103 (2009)
26. Mamaev, M.A, Petrenko, S.A.: Technologies of Information Protection on the Internet, 848 p. Publishing house "Peter", St. Petersburg (2002)
27. Markov, A.S., Tsirlov, V.L., Barabanov, A.V.: Methods for Assessing the Discrepancy Between Information Protection Means; [ed. A. S. Markov], 192 p. Radio and communication, Moscow (2012)
28. Markov, A.S.: Chronicles of cyberwar and the greatest redistribution of wealth in history. Quest. Cybersecurity. **1**(14), 68–74 (2016)
29. Petrenko, A.A., Petrenko, S.A.: Cyber units: methodical recommendations of ENISA. Quest. Cybersecurity. **3**(11), 2–14 (2015)
30. Petrenko, A.A., Petrenko, S.A.: Intranet Security Audit (Information Technologies for Engineers), 416 p. DMK Press, Moscow (2002)
31. Petrenko, A.A., Petrenko, S.A.: Research and Development Agency DARPA in the field of cybersecurity. Quest. Cybersecurity. **4**(12), 2–22 (2015)
32. Petrenko, A.A., Petrenko, S.A.: The way to increase the stability of LTE-network in the conditions of destructive cyber-attacks. Quest. Cybersecurity. **2**(10), 36–42 (2015)
33. Petrenko, A.S., Petrenko, S.A.: Super-productive monitoring centers for security threats. Part 2. Protect. Inf. Inside. **3**(75), 48–57 (2017)
34. Petrenko, A.S., Petrenko, S.A.: The first interstate cyber-training of the CIS countries: "Cyber-Antiterror2016". Inf. Protect. Inside. **5**(71), 57–63 (2016)
35. Petrenko, S.A.: Methods of ensuring the stability of the functioning of cyber systems under conditions of destructive effects. Proceedings of the ISA RAS. Risk Manag. Security, **52**, 106–151 (2010)
36. Petrenko, S.A.: Methods of Information and Technical Impact on Cyber Systems and Possible Countermeasures. Proceedings of ISA RAS. Risk Manag. Security, **41**, 104–146 (2009)
37. Petrenko, S.A., Kurbatov, V.A., Bugaev, I.A., Petrenko, A.S.: Cognitive system of early warning about computer attack. Protect. Inf. Inside. **3**(69), 74–82 (2016)
38. Petrenko, S.A., Petrenko, A.A.: Ontology of cyber-security of self-healing SmartGrid. Protect. Inf. Inside. **2**(68), 12–24 (2016)
39. Petrenko, S.A., Petrenko, A.S.: Creation of a cognitive supercomputer for the computer attacks prevention. Protect Inf. Inside. **3**(75), 14–22 (2017)
40. Petrenko, S.A., Petrenko, A.S.: From detection to prevention: trends and prospects of development of situational centers in the Russian Federation. Intellect Technol. **1**(12), 68–71 (2017)
41. Petrenko, S.A., Petrenko, A.S.: Lecture 12. Perspective tasks of information security. Intelligent information radiophysical systems. Introductory lectures [A. O. Armyakov and others; ed. S.F. Boev, D.D. Stupin, A.A. Kochkarova], pp. 155–166. MSTU them. N.E. Bauman, Moscow (2016)

42. Petrenko, S.A., Petrenko, A.S.: New doctrine as an impulse for the development of domestic information security technologies. Intellect Technol. **2**(13), 70–75 (2017)

43. Petrenko, S.A., Petrenko, A.S.: New doctrine of information security of the Russian Federation. Inf. Protect. Inside. **1**(73), 33–39 (2017)

44. Petrenko, S.A., Petrenko, A.S.: Practice of application of GOST R IEC 61508. Inf. Protect. Insider. **2**(68), 42–49 (2016)

45. Petrenko, S.A., Shamsutdinov, T.I., Petrenko, A.S.: Scientific and technical problems of development of situational centers in the Russian Federation. Inf. Protect. Inside. **6**(72), 37–43 (2016)

46. Petrenko, S.A., Simonov, S.V.: Management of Information Risks. Economically Justified Safety (Information technology for engineers), 384 p. DMK-Press, Moscow (2004)

47. Petrenko, S.A.: The concept of maintaining the efficiency of cyber system in the context of information and technical impacts. Proceedings of the ISA RAS. Risk Manag. Safety. **41**, 175–193 (2009)

48. Petrenko, S.A.: The Cyber Threat model on innovation analytics DARPA. Trudy SPII RAN. **39**, 26–41 (2015)

49. Petrenko, S.A.: The problem of the stability of the functioning of cyber systems under the conditions of destructive effects. Proceedings of the ISA RAS. Risk Manag. Security. **52**, 68–105 (2010)

50. Petrenko, S.A., Kurbatov, V.A.: Information Security Policies (Information Technologies for Engineers), 400 p. DMK Press, Moscow (2005)

51. Petrenko, S.A.: Methods of detecting intrusions and anomalies of the functioning of cyber system, Proceedings of ISA RAS. Risk Manag. Safety. **41**, 194–202 (2009)

52. About formal bases of OWL [Electronic resource]. Access mode: http://semanticfuture.net/index.php. Accessed 20 Dec 2014

53. Ashby, U.R.: Principles of Self-Organization, pp. 314–343. Mir, Moscow (1966)

54. Bongard, M.M.: The Problem of Recognition. Fizmatgiz, Moscow (1967)

Chapter 3
Limitations of Von Neumann Architecture

3.1 Creation of a Super-high Performance Supercomputer

As part of its strategic goals for technological development, the Russian Federation aims to create ultrahigh productivity supercomputer technologies or exascale computing by 2025. To this end, it is necessary to effectively develop domestic production of highly productive and trusted computer aids; to overcome physical limitations relating to energy consumption, reliability, and structural dimension of modern processors; and also to develop and implement effective organization of exascale calculation. Intensive research work is taking place in Russian research and education institutions to solve these scientific and technical problems [1–5].

3.1.1 Problem Overview

In several technologically advanced countries (the USA, China, Germany, Japan, the UK, France, Russia, South Korea, etc.), supercomputer technologies are being rapidly developed. For example, "Moore's law" dictates that every 18 months, the performance of modern computing systems doubles, which means increases of 1000 times for every 11 years (see Inset 3.1).

Inset 3.1 Computer Performance Dynamics Growth

$\rightarrow$MFLOPS (**1975**, 10^6) $\rightarrow$ GFLOPS (1986, 10^9) $\rightarrow$ TFLOPS(1997, 10^{12})
$\rightarrow$PFLOPS (**2008**, 10^{15}) $\rightarrow$ EFLOPS (2019, 10^{18}) $\rightarrow$ ZFLOPS (2030, 10^{21})

© Springer International Publishing AG, part of Springer Nature 2018
S. Petrenko, *Big Data Technologies for Monitoring of Computer Security: A Case Study of the Russian Federation*, https://doi.org/10.1007/978-3-319-79036-7_3

Annually, different world rankings (e.g., TOP 500[1]) are compiled to evaluate the level of supercomputer technologies and determine the winners among supercomputers based on their performance (excluding high-security facility informatization computers). On these terms, the best domestic supercomputer from Moscow State University took 52nd place on the Linpack testing task in the 48th TOP 500[2] rating with a PFLOPS real performance value of 2102. The leading *Sunway TaihuLight* supercomputer, located in the *National Supercomputing Center in Wuxi China*, demonstrated 93 PFLOPS (with a peak performance of 125,43 PFLOPS).

In this sense, supercomputer performance is understood as the number of floating-point instructions conducted per second. In this case, a distinction is made between peak performance as a maximum possible (idealized) number of instructions per second and real performance on a given task with a real number of instructions per second required to solve the computing task. The ratio of the real performance to the peak is called the efficiency of the supercomputer on a given problem. Clearly, supercomputer efficiency can vary on various computing tasks [4, 6–8]. The indicator value of real performance on testing tasks with a large number of unknowns (Linpack) are used in the TOP500 supercomputers world rating for comparing computer-based system performance. There are also other tests [4, 6–8]: for example, in the Graph500[3] ranking supercomputer real performance is calculated on the testing task of searching the breadth of a large graph (from the class of tasks requiring intensive data processing).

A critical analysis of the TOP500 published results demonstrated that supercomputers have significant structural (Figs. 3.1 and 3.2) and functional diversity (Table 3.1) including a significant spread of the values of both real and peak performance not only among the first hundred but also among the ten highest-rated computers (Table 3.2).

3.1.2 Relevance of the Problem

The relevance of organizing exascale calculation owes to the need to solve time-consuming computational problems in the strategically critical public domain (Table 3.3) with the required accuracy and correctness. Indeed, today high-performance computing systems for solving problems requiring more than 10^{20} computational operations are in high demand (for comparison, the number of atoms in the universe is estimated at 10^{80}). Examples of such problems are given below.

---

[1] https://www.top500.org/

[2] https://www.top500.org/lists/2016/11/

[3] https://www.Graph500.org/

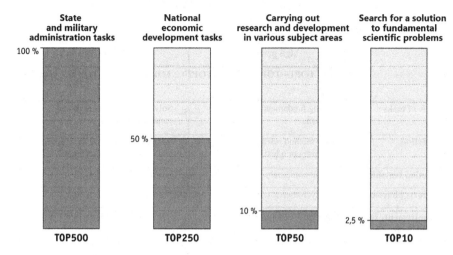

Fig. 3.1 Supercomputer differentiation in TOP

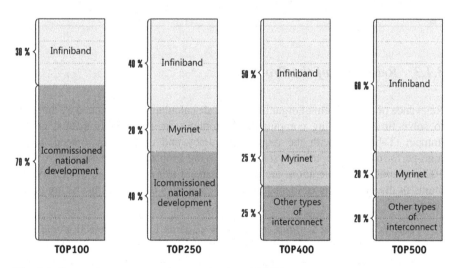

Fig. 3.2 Types of supercomputer interconnect from TOP500

A critical analysis of the TOP500 published results demonstrated that supercomputers have the significant structural (Figs. 3.1 and 3.2) and functional diversity (Table 3.1) including a significant spread of the both real and peak values of performance not only in the first hundred but also in the top of ten rating (Table 3.2).

Table 3.1 Differences among TOP500 supercomputers

	Computing system with ultra-high performance		Computing system with high performance
	TOP1←TOP10	←TOP20...TOP100→	TOP250→TOP500
Scale of tasks	Solving fundamental scientific problems		Solving applied research and technology tasks
Direction	Multi-method research studies of immense complexity		Subject research and development and PAC prototype design
Solution scalability	Special computers		Universal computers
Imports phase-out	Mainly domestic components		Mainly foreign components
Requirements for components production	Requires a unique development		Available ready-to-go solution
Financing	Governmental		Governmental and large business
Location	Large national supercomputer centers		National regional and corporate supercomputer centers

3.1.3 Relevance of the Problem

Relevance of the problem of organizing exascale calculation is explained by the need to solve time-consuming computational problems in strategically critical public domain (Table 3.3) with the required accuracy and correctness. Indeed, today high-performance computing systems for solving problems that require more than 1020 computational operations are in high demand (for comparison, the number of atoms in the universe is estimated at 1080). Examples of such problems are given further.

1. Synthesis of mass and group cyber-attack patterns for early detection and prevention of cyber-attacks based on known methods of computational mathematics and computer modeling, including digital signal processing and multichannel digital filtering, as well as calculating possible "trajectories" of cyber-attacks and synthesizing the required warning scenarios based on the Monte Carlo method [4, 6–8]. In this work, exaFLOP supercomputer performance is required not for directly increasing the number of computational operations but for interface computation of multi-scale confrontation processes in cyberspace with given accuracy in real time and also for the most effective approximations and algorithms for solving various classes of information security problems.

2. Computer modeling of the brain and integration into the network architecture of a large number of neurons (several million or more) linked by synaptic connections. The solution to this problem is relevant for the design of the hierarchical

Table 3.2 Supercomputer classification TOP500

Class	Rating position	Description of supercomputer class	Number of microprocessor cores, no more than, in mil.	Maximum real performance, PFLOPS	Maximum peak performance, PFLOPS	Power consumption, no more than, MW
I	1–7 (7 supercomputers)	National supercomputer centers	10.7	100	150	20
II	8–117 (110 supercomputers)	Regional and sector-specific supercomputer centers	1.4	10	30	15
III	118–316 (199 supercomputers)	Corporate supercomputer centers	0.2	5	10	10
IV	317–500 (184 supercomputers)	Supercomputer centers of individual enterprises and research laboratory	0.3	1	5	5

Table 3.3 Expert rating of capacity requirement

Subject area	2018	2022	2025
National security	Max. 100 PFLOPS	800 PFLOPS	10–30 EFLOPS
Aerospace defense	Max. 75 PFLOPS	800 PFLOPS	10–25 EFLOPS
Information security	Max 50 PFLOPS	750 PFLOPS	10–20 EFLOPS
Mechanical engineering automobile manufacturing	20 PFLOPS	400 PFLOPS	5–20 EFLOPS
Aircraft industry	40 PFLOPS	500 PFLOPS	5–30 EFLOPS
Shipbuilding	30 PFLOPS	500PFLOPS	5–20 EFLOPS
Rocket engineering	50 PFLOPS	750 PFLOPS	5–30 EFLOPS
Nuclear engineering	Max. 100 PFLOPS	900 PFLOPS	10–30 EFLOPS
Oil and gas industry	40 PFLOPS	700 PFLOPS	10–25 EFLOPS
Material engineering	30 PFLOPS	450 PFLOPS	5–20 EFLOPS
Biotechnology	25 PFLOPS	450 PFLOPS	5–25 EFLOPS
Public health service	25 PFLOPS	500 PFLOPS	5–20 EFLOPS
Research and education	50 PFLOPS	750 PFLOPS	10–30 EFLOPS

system of the "solvers" (control units) of the national cyber-attack early-warning system [4, 6–8].

3. Calculation of the combat maneuvering aircraft aerodynamic characteristics, which is characterized by flights with large angles of attack, based on direct numerical simulation turbulence models and unsteady calculation methods [4, 6–8]. Calculation of the air weapons (AW) jet impact on the aircraft details at position DNS, taking into account air weapon moving and final combustion AW jet.

4. Calculation of the missile-carrying aircraft aerodynamic characteristics, modeling work process in turbo-pumping unit, fuel and oxidizing material of liquid rocket engine, and modeling work process in combustion chamber of liquid rocket engine [4, 6–8]. Including modeling combustion process in engines with new designs (from the course of a reaction at the individual molecules level to vortex formation with introducing fuel into combustion chamber) based on quantum molecular dynamics and classical molecular dynamics models and methods, kinetic Monte Carlo method large eddy simulation, direct numerical simulation based on Navier-Stokes system, Reynolds time-averaged equation models, and others.

5. Complex physical modeling of fourth-generation nuclear reactors with fast neutron and fuel cycle closure linked processes in 3D formulation: heat hydraulics, neutron and gamma transfer, radiation safety, and simulation of material structure behavior [4, 6–8]. Using direct numerical simulation of thermophysical processes and material properties from the first principles, taking into account the molecular dynamics interaction.

6. Dynamic 3D analysis of nominal and transient reactor condition in the full cluster approximation, taking into account the isotope-isometric fuel kinetic and fission product; multidimensional, multiphase thermohydraulics of the first and the second circuits; turbulent mixing; discrete approximation of dispersed particles; and complex hydraulic network geometry [4, 6–8].

7. In active zone neutron-nuclear processes modeling based on anisotropic kinetic approximation methods for the numerical reasons of the atomic power station design and safety [4, 6–8]. The computation time on a supercomputer with a performance of 1 PFLOPS will be 1,000,000 hours and 1000 h on a supercomputer with a performance of 1 EFLOPS.

In technologically developed countries across the world including the Russian Federation, a course has been taken to create a national supercomputer with exaFLOP performance that is 10^{18} to solve the tasks mentioned above and many other computing problems which are highly intensive in terms of labor and complexity. In this context, supercomputers with exaFLOP performance are computation systems whose functional characteristics outstrip the majority of fifth-generation computers: more operations per second, bigger memory capacity and high rates of internal data exchange (bandwidth, latency, message rate).

3.1.4   Development Programs

A set of priority action items called the exascale initiative was developed in 2009 with the explicit aim of preserving the USA's leading role in the field of supercomputer technologies. Under this program, a number of research programs aiming to achieve exaFLOP performance by 2019–2020 (Table 3.4) including DAPRA program were initiated:

- *Omnipresent High Performance Computing*, from 2011 to the present day
- *Semiconductor Technology Advanced Research Network*, STARTnet, from 2013 to the present day
- *Ubiquitous Hign Performance Computing*, UHPC, from 2013 to the present day
- *High Productivity Computing Systems*, HPCS, 2002–2013, and others

Under the DAPRA program, a number of research centers were opened[4]:

---

[4]http://www.sdsc.edu/pmac/papers/docs/dongarra2008darpa.pdf

Table 3.4 Development programs of supercomputer technologies

Plans	Funding and research results		
	USA	EU	PRC
Known development program of supercomputer technologies	DOE – more than 100 million dollars/year NSF and universities – up to 30 million dollars/year DAPRA UHPC – about 75 million dollars for 4 years In total, more than 650 million dollars/year is allocated for research	European Commission, PRACE, DEISA and EESI programs, national programs, universities, and industry. A total of 200 million euros are allocated for research study	**2015:** petascale supercomputer based on own hardware components. The open funding part is over 500 million dollars/year
Up to 10 PFLOPS	**2011–2012:** 20–34 PLOPS Sequoria and titan 10 PFLOPS – Mira	**2012–2013:** SuperMUC – 3PFLOPS; Hermit - 4–5 PFLOPS	**2015:** Supercomputer Tianhe-2 at National Super Computer Center in Guangzhou with 33.863 PFLOPS performance
Up to 100 PFLOPS	**2015–2016:** 50 supercomputers (from 0.833 to 17.5 PFLOPS) 2016: In DOE/SC/Oak Ridge National Laboratory Titan supercomputer with 17.590 PFLOPS performance	**2016:** the best supercomputer is Piz Daint from National supercomputer center of Swizerland – CSCS 9.779 PFLOPS	**2016:** supercomputer Sunway TaihuLight at National Supercomputing Centre in Wuxi China with performance 93.015 PFLOPS
1 EFLOPS – 10 EFLOPS	**2020:** up to 10 supercomputers with 1–10 EFLOPS performance	**2020:** up to 5 supercomputers with 1–10 EFLOPS performance	**2020:** up to 10 supercomputers with 1–10 EFLOPS performance

- *The Center for Future Architectures Research, C-FAR* at University of Michigan
- *Spintronic Materials, Interfaces and Novel Architectures, C-SPIN* at University of Minnesota
- The Center for Function Accelerated Nano Material Engineering, FAME at University of California, Los Angeles
- *The Center for Low Energy Systems Technology, LEAST* at University of Notre Dame du Lac
- *The Center for Systems on Nanoscale Information Fabrics, SONIC* at University of Illinois
- *Terra Swarm Research Center*, TerraSwarm at University of California, Berkeley and others

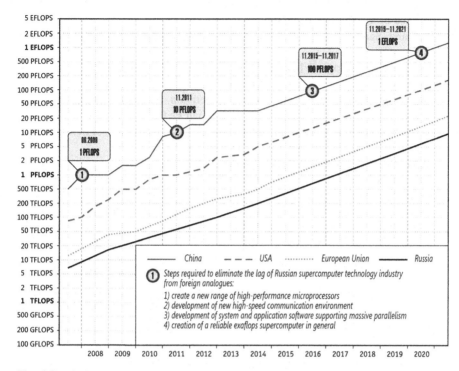

Fig. 3.3 Priority objectives and tasks

Currently, over 40 leading USA universities are taking part in the DAPRA programs, and around 200 doctors and more than 500 postgraduates are conducting scientific research studies. The *US Air Force Research Laboratory* (*AFRL*) is acting as the commissioning party on behalf of the US Department of Defense. *Semiconductor Industry Association* (*SIA*) and major computer equipment manufacturers such as Applied Materials, IBM, Intel Corporation, Micron Technology, Raytheon, Texas Instruments, and United Technologies are also taking part in the research.

In the Russian Federation, the problem of creating a supercomputer with exaFLOP performance (Fig. 3.3) is the object of several Federal Target Programs (FTPs):

- "Research and developments in priority growth areas of Russian science and technology sector for the period of 2014–2020": coordinator, Ministry of Education and Science of the Russian Federation; governmental customers, Ministry of Education and Science of the Russian Federation and Lomonosov Moscow State University
- "Information society for 2012–2020": coordinator, Ministry of Communications and Mass Communications of the Russian Federation
- "Research and developments in priority growth areas of Russian science and technology sector for 2007–2011": coordinator, Ministry of Education and

Science of the Russian Federation; governmental customers, Ministry of Education and Science of the Russian Federation and Lomonosov Moscow State University

- "National technological base for 2007–2011 years": coordinator, Ministry of Industry and Trade of the Russian Federation; governmental customers, Ministry of Industry and Trade of the Russian Federation, Ministry of Education and Science of the Russian Federation, The Russian Federal Space Agency, Russian Academy of Sciences, Siberian Branch of the Russian Academy of Sciences, and State Nuclear Power Corporation "Rosatom"
- The list of projects in accordance with the "Development of supercomputers and grid technologies" (2011) strategy, approved by the Russian presidential commission for modernization and technological development of Russia's economy; governmental customer, State Nuclear Power Corporation "Rosatom"; general contractor, FGUP Russian Federal Nuclear Centre All-Soviet Union Scientific research institute of experimental physics
- "Development of electronic component base and radio electronics for 2008–2015": coordinator, Ministry of Industry and Trade of the Russian Federation; general contractor, Ministry of Industry and Trade of the Russian Federation, Ministry of Education and Science of the Russian Federation, The Russian Federal Space Agency, and State Nuclear Energy Corporation "Rosatom"

Immediacy of the problem of exaFLOP calculation organization is confirmed by requirements of the following normative legal documents:

- The Foreign Policy Concept of the Russian Federation (approved by the Decree of the Russian Federation President #640 of November 30, 2016)
- The National Security Strategy of the Russian Federation (approved by the Russian Federation Presidential Decree #683 of December 31, 2015)
- The Doctrine of Russian Federation Information Security (approved by the Russian Federation Presidential Decree # 646 of December 5, 2016)
- Russian Federation Presidential Decree of July 7, 2011 # 899 about "A Confirmation of priority of the development of science, technologies and engineering area in the Russian Federation and a range of critical Russian Federation technologies"
- Development strategy of information technologies in the Russian Federation for 2014–2020 and long term until 2025
- The concept of the federal program "Development of high-performance computing technologies based on exascale supercomputer (2012–2020)," State Nuclear Power Corporation "Rosatom," 2011
- Project of realization the integral technological platform "National supercomputer technological platform," Lomonosov Moscow State University, 2011
- Development strategy of supercomputer production
- Project "Development of supercomputer and grid technologies," 2010–2012
- Russian Federation Presidential Decree # 648 of July 25, 2013 "On building the system of distributed situation centers, operating under the single liaison protocol"

- Program for modernization of existing and creation of new situation centers, in execution of an order of the Prime Minister of the Russian Federation D.A. Medvedev, # DM-P7–5840 of August 24, 2015
- Russian Federation Presidential Decree of January15, 2013 "On creation of governmental system for detection, prevention and control of cyber-attacks on the information resources of the Russian Federation"
- The concept of the state detection, prevention, and control system of cyber-attacks on the information resources of the Russian Federation (approved by the President of the Russian Federation of December 12, 2014, № K 1274), etc.

The president of the US Council on Competitiveness Deborah Wince-Smith at the annual conference opening on HPC Conference in 2004 clarified: "The country that wants to outcompete (author's note: dominate in cyber, marine, space and air space) must outcompute." As S.M. Abramov, Director of the Program Systems Institute, rightly noted: "Supercomputer technology is the key critical technology and the only tool that gives an opportunity to beat the competition." It does not matter in which economic sector the competition goes, the citation pertains to the manufacturing, mining, and processing industries and also for the purposes of state and military administration, national security, and developing technological innovations in cybersecurity. In this way, the immediacy of developing exaFLOPS calculation is practically demonstrated as well as its significant academic and applied relevance.

3.1.5 Expected Results

The national program for the creation and development of high-performance super-computer technologies [1, 2, 5, 6, 9, 10] projects the development of a supercomputer with the following characteristics: a real performance of at least 1–3 EFLOPS, high-capacity data exchange from 10^{10} bytes/s, and power consumption no more than 10–20 MW.

To achieve these results, the following actions are required:

- To design a domestic exaFLOP supercomputer which allows effective execution of applications from 10^8 cores
- To develop a basic set of mentioned supercomputers (experimental system), capable of verifying the design solution
- To create reliable hardware components for exaFLOP supercomputer, which comply with design constraints and reliability requirements
- To create the corresponding system software and application software for providing exaFLOP computations

Besides that, it is necessary to develop a domestic series of reliable MIMD/SIMD processors with a performance of at least:

- 64/500 GFLOPS (design rules – 65/30 nm)
- 144/200 GFLOPS (design rules – 45/22 nm)

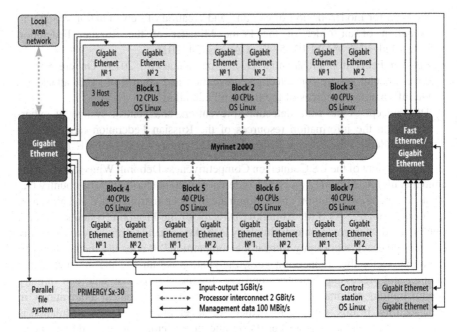

Fig. 3.4 Functional scheme MBC 6000 IM

- 2000/16000 GFLOPS (design rules – 22/17 nm)

The domestic supercomputer's power consumption should not exceed 300–500 W. For comparison, Intel is planning to achieve the performance target 8 nm by 2018 and to create a series of MIMD/SIMD processors with the following performance characteristics:

- 256–512/1500–2000 GFLOPS (design rules – 30/30 nm)
- 500–1000/4000–8000 GFLOPS (design rules – 22/22 nm)
- 1000–2000/10,000–16,000 GFLOPS (design rules – 17/17 nm), respectively

At the same time, the Russian processor architecture (Fig. 3.4) should include:

- Superfluous duplicate cores, which, in the event of failure, are to replace the broken ones, keeping topology and providing low-level reliability growth depending on the redundancy rate (by analogy with memory corrective).
- Means for dynamic reconfiguration of processor organization, providing core division of MIMD and SIMD components into specific dovetail into other subsets (interconnected), the structure and performance of which are determined in accordance with the parameters of the executable process.

Table 3.5 Required characteristics

Parameter	2017	2020
Supercomputer peak performance	100 PFLOPS	1–3 EFLOPS
Number of MIMD/SIMD cores, pcs	10^7	10^8
Power consumption, MW	5	20
Main-memory capacity, PB	5	50
Peak performance of MIMD/SIMD processor, GFLOPS	8000	16,000
Peak performance of MIMD/SIMD processor computation module, GFLOPS	5000	10,000
Overall service capacity of computation module interface, GB/S	2500	25,000
Number of MIMD/SIMD processor, pcs	25,000	100,000
Number of computation modules, pcs	3000	10,000
External memory of parallel FS, PB	100	1000
Data rate with external memory, TB/S	10	100

Also, it is essential to provide the required processor interconnect speed between processors, computation module, their clusters, and between racks ("dragonfly – environments") at levels (copper, optical fiber) from 10^{10} b/s to 10^{12} b/s. Finally, it is necessary to develop input-output system which can provide a number of parallel I/O streams 10^8; number of files is about 10^{16}–10^{17} bytes. For achieving such characteristics, a number of advanced research areas and developments have been planned in the Federal program framework "Development of high-performance calculation technologies based on exascale supercomputer (2012–2020)" [4, 6–8].

With regard to *multiprocessor computing system development* (Table 3.5):

- Applying hybrid architecture and applying arithmetic accelerators (GPU, FPGA, and other specific hardware features)
- Expanding the number of processor cores both in the whole system and inside the central processing element
- Increasing communication environment complexity
- Providing a speed of transmission at multiple levels

With regard to multiprocessor environment improvement:

- Development for various execution algorithm topologies of data exchange operation between compute-processing elements and also between multiprocessor environment and external memory; development of clustered environment builder, in which processor elements are also environments
- Study of routing and communication capabilities for various multiprocessor environment topologies
- Estimation of multiprocessor environment efficiency on the exchange time complexity (including global) for various topologies and data exchange mechanisms

- Development of conflict-free exchange sets builder for environment with different connection topology, estimation of sets quantity, and power

With regard to *hybrid computing architectural design*:

- Study of acceleration magnitude calculation by hybrid systems and their efficiency, taking into account the operation of Amdahl's and Gustafson's laws (division and multiplying methods) by 10^8 cores and generalization for recursive hybrid architectures
- Study of hybrid systems efficiency scaling engine in conditions of their complexity increasing
- Study of architectural reliability engineering of hybrid computing system
- Study of interaction between exascale computing environment with external memory and graphics subsystem

With regard to the *development of mathware and software*:

- Introduction of newly improved (more expensive from a computational point of view) design schemes and approximations (in addition to creating new ones and modifying existing computational algorithms, numerical studies of the accuracy, and justification of new algorithms, as well as an analysis of their applicability to the classes of simulation problems under consideration)
- Modification of existing and creation of new mathematical techniques, due to the general increase of computational cores, with goal of efficient parallelization for tens and hundreds of millions of processes and orientation to mixed multilevel algorithms, for example, using the parallel memory model (MPI) between separate elements and shared memory models (OpenMP tools) inside the nodes
- Adaptation of mathematical techniques to the next generation computing systems, including the transition of mathematical techniques to the effective use of hybrid architecture (currently, in this area of computer technology, established standards of hardware and software have not yet been formed, and there is not significant experience in using these tools)
- Development of scalable programming tools that implement the parallelization and simultaneous execution of $\sim 10^8$ processes
- Creation of a unified system for task management, resources of computer complexes and operational monitoring that optimize the process location by computational elements and implement job queue management algorithms that take into account their priority, as well as temporal, structural parameters among others
- Development of tools for collecting, accumulating, and analyzing statistical information on the user program account, as well as collecting and analyzing the hardware and software components of computer systems status (error statistics, equipment failures, routine, and emergency operations)

3.2 Development Program for Supercomputer Technologies

Whether exaFLOPS is a super ambition or an unavoidable barrier for further economic development is a pressing question for modern supercomputer technologies development throughout the world. In 2011, specialists from FGUP All-Russian Research Institute for Experimental Physics and the Russian Academy of Sciences and Industrial Enterprises developed the concept of creating a domestic exascale computer.

At the end of 2016, summarizing the intermediate results, V.B. Betelin from the Russian Academy of Sciences and representatives of OJSC "T-Platform" noted that for the previous 5 years, the basic principles articulated in the concept were justified: supercomputers had developed toward application of the hybrid schemes along with universal processors and vector accelerators in the architecture. At the same time, related research continued in national schools of sciences, despite the forecast of slower development in other countries. The USA, China, and the EU had planned to create exaflops by 2018; however, now the target year is around 2022–2023 [2, 3, 11–15].

3.2.1 Existing Capacity

The national development program of the supercomputer technologies began in 1996, when the RAS Interdepartmental Supercomputer Center was established in Moscow to implement the programs of the Ministry of Education and Science and the Russian Foundation of Fundamental Research.

Since then, Russia and Belarus have successfully completed a number of the following programs:

- *SKIF* (2000–2004) – "Development and conversion of serial production of the high-performance computing system models with parallel architecture (supercomputers) and creation of applied hardware and software complexes based on them"
- *TRIADA* (2005–2008) – "Development and implementation of knowledge-intensive computer technologies based on multi-processor system in the Union member states"
- *SKIF-GRID* (2007–2012) – "Development and application of grid-technologies soft hardware and advanced high-efficiency (supercomputer) computer systems of 'SKIF' family" (Table 3.6)

Table 3.6 Supercomputer of "SKIF" family characteristics

№	Years, peak performance (estimated range)	Cores in CPU/precision	Network solutions	Form factor (CPUs/U)	Notes
1	2000–2003, 20–500 GFLOPS	1/32	FastEthernet/ SCI(2D-top), Myrinet	4 U–1 U (0.5–2)	National SCI (2D-top). Cooldown: air
2	2003–2007, 0.1–5 TFLOPS	1/32–64	GbEthernat/SCI (3D-top), Infiniband	1 U, HyberBlade (2)	ServNet v.1, v2. Accelerator: FPGA, OBC.Cooldown: air
3	2007–2008, 5–150 TFLOPS	2–4/64	GbEthernet/ Infinitband DDR	1 U, blades 20 CPU in 5 U (2–4)	ServNet v.3. Cool-down: air-water-freon
4	2009–2012, 0.5–5 PFLOPS	4–12/64	Infinitband QDR/national system area net-work (3D-top)	Blades 64 CPU in 6 U (10.667)	ServNet v.4 new approaches to cool-down. Accelerator: FPGA, GPU, MSOS

At the meeting of the Russian Federation Security Council, devoted to the development of strategic information technologies in July 2009, D.A. Medvedev set a task to overcome Russia's lagging pace in supercomputer production. He observed: "Russia should invest in the development of supercomputer technologies not only because it is a trendy topic. but also because otherwise we can't produce competitive products, which will be favorably perceived by potential customers." Further, the Security Council of the Russian Federation adopted the *Supercomputer Production Development Strategy.*

In 2011 the presidential program *Development of supercomputers and GRID-technologies* was adopted. The program initiated the creation and development of more than 20 regional supercomputer centers (Table 3.7) located in large national scientific centers. Plans were also made to create about 40 new industry supercomputer centers in companies and in Research Institute knowledge-intensive industries.

At this stage, a number of enterprises and agencies described briefly below are part of the collaboration between science and industry in order to establish and develop ultrahigh performance supercomputer technologies.

3.2.2   JSCC RAS

Joint Supercomputer Center of the Russian Academy of Sciences (JSCC RAS)[5] *provides research and methodology support* for research on national supercomputer

---

[5]http://www.jscc.ru/

Table 3.7 Characteristics of some national supercomputers

Rating position in TOP 50	Basic peak performance, TFLOPS	Basic core amount	Location
1	1700.21	82,468	Lomonosov Moscow State University (Moscow)
4	236.82	14,016	South Ural State University (Chelyabinsk)
6	216.56	1192	N.N. Krasovskii Institute of Mathematics and Mechanics of the Ural branch of the Russian Academy of Sciences (Ekaterinburg)
13	103.31	3040	National Research Lobachevsky State University of Nizhni Novgorod
15	62.35	5424	National Research Tomsk State University
21	85.69	2400	Siberian supercomputer center of SB RAS (Novosibirsk)
31	22.702	3520	Institute of system dynamics and control science (Irkutsk)
35	23.50	1920	North-Eastern Federal University (Yakutsk)
36	32.74	980	Belgorod National Research University
37	40.14	1536	Saint Petersburg State University
43	21.12	2268	Ufa State Aviation Technical University
45	18.84	2072	Taganrog technological university of southern federal region
46	17.89	1920	Vyatka State University (Kirov)
48	16.87	1808	Siberian Federal University (Krasnoyarsk)

technologies development and also provides scientists and practicians with the required computing resources (Fig. 3.5). It includes MVS -10P and MVS -100K.

MVS -10P with performance of 375.7 TFLOPS (459 rating position in TOP 500 as of November 2016) (peak performance is 523.8 TFLOPS) contains 207 computation nodes, each of which consists of two *Intel Xeon E5-2690* processors with 64GB RAM, as well as two *Intel Xeon Phi Z110X* co-processors. Computation nodes are integrated into a traffic system based on FDR Infiniband and a control network based on Gigabit Ethernet. There are two task management systems: parallel task management system and SLURM.

MVS -100K with performance of 119.93 TFLOPS (peak performance is 227.94 TFLOPS) contains 1275 modules, each is equipped with two four-core/six-core *Intel Xeon processors*. Computation nodes are integrated into a generalized definitive field based on Infiniband technology.

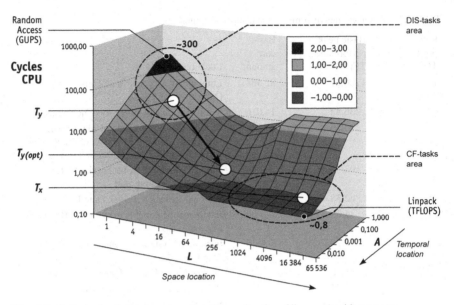

Fig. 3.5 Optimization technology based on the results of profiling work with memory

3.2.3 National Research Center "Kurchatov Institute"

National Research Center "Kurchatov Institute"[6] holds the convergent research study on the NBIC technologies (nano, bio, info, and cognitive) and also finds solutions to the following problems:

- Process modeling for nuclear power stations and nuclear submarines and engineering design for creating new materials based on nano- and biotechnologies (applying the supercomputers of the abovementioned center – 462nd in TOP500 as at November 2016), which are firmly coupled clusters with performance up to 374 TFLOPS, consolidated with data network based on InfiniBand
- Data operation from mega machine and also data from international experiments – XFEL, FAIR, ITER (applying loosely coupled center clusters, petabyte storage array, and associated software)

---

[6]http://www.nrcki.ru/

3.2.4   RFNC Computer Center

The Center[7] carries out its activities for the benefit of the nuclear weapons complex, military-industrial complex, and other industrial enterprises of the Russian Federation. For example, copyright software packages such as LOGOS, DANKO +GEPARD, NIMFA, TDMCC, CONCORD, SERENA, FIRECON, MeltCup, etc. are used to solve a wide range of complex physical process simulation modeling tasks.

The computational center of RFNC-VNIIEF includes:

- Network infrastructure – hardware-software complex, designed to provide access to the computing resources as well as to provide a secure channel
- Basic set of computational systems – multiprocessor supercomputers of different classes, designed to implement computational analysis and complex full-scale simulation modeling of complicated engineering system and physics processes
- System software that consists of Linux family OS, communication, and other software for implementation of high-performance parallel computing
- Archive system – firmware for the file data storage and data backup
- Visualization system- firmware for the graphic processing and analysis of large data volumes resulting from computational analysis

3.2.5   Research Institute for System Studies

The Federal government agency "Scientific Research Institute of System Development in Russian Academy of Sciences" (NIISI RAN)[8] carries out research on the development of microprocessing and communication VLSIs, optimal for solving the designated class of problem, including the development of microprocessors with COMDIV architecture (Fig. 3.6).

For instance, the microprocessor for digital signal processing 1890BM9Y (2016) consists of:

- Two universal 64-bit cores, each of them includes four computational sections of single-precision floating-point arithmetic
- Four switched channels in the RapidIO 4x standard with a transfer speed of up to 3.125 Gbit/s

---

[7]http://www.vniief.ru/
[8]https://www.niisi.ru/

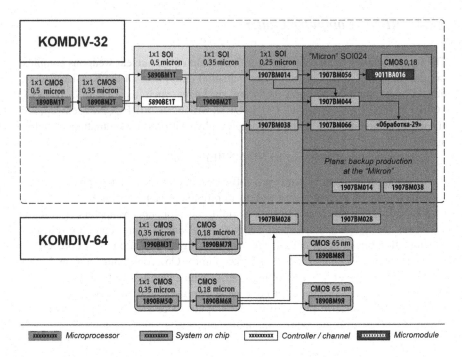

Fig. 3.6 KOMDIV microprocessors

- Two DDR2/3 controllers
- Two 1Gb/s Ethernet controllers
- SATA 3.0 controller
- USB 2.0 host-controller, etc.

3.2.6 Moscow Center of SPARC Technologies (MCST)

MCST[9] was set up in 1992 based on Lebedev Institute of Precise Mechanics and
Computer Engineering (*IPMCE*) where in the 1970s and 1980s the well-known
MVK *Elbrus-1* and *Elbrus-2* were developed under the guidance of V.S. Burtsev.

The company has developed four generations of SPARC-compatible processors:
MTST R1000; *Elbrus-2C +*, with a frequency of 500 MHz and two cores; *Elbrus-4S*
with a frequency of 800 MHz and four cores; and *Elbrus-8S* (designed according to
the 28 nm technology) with eight cores. In 2015, the development was completed for
a single-core microprocessor *Elbrus -1S+* with integrated graphics engine,
supporting 3D graphics hardware with support for hardware acceleration of 3D
graphics and a total power consumption of less than 10 W. Additionally, an external

---

[9]https://www.mcst.ru/

peripheral interface processor, the "south bridge," a microcircuit chip-concentrator (2011), linked peripheral devices, and bus with a central processor were developed. In 2015 the development of KPI-2 with a bandwidth capacity of at least 16 Gbyte/s (8Gbyte/s at each direction) was completed. The KPI-2 has a PCI Express 2.0 controller for 20 bus, 3 Gigabit Ethernet controllers, and 8 SATA 3.0 controllers. Most of the JSC "MCST" processor models permitted integration into multiprocessor machines, up to four processors in the system, and each processor could have its own "south bridge."

A wide range of computing systems of various classes was developed on the basis of "Elbrus" processors: from desktop computers and laptops to servers and built-in appliances, including those for severe operating conditions. "Elbrus" OS is based on the Linux 3.10 core (2.6.33). The "Elbrus" OS default distribution is Debian. A number of optimizing compilers for C, C ++, Fortran languages have been developed.

3.2.7 Institute of Multiprocessor Computing Systems

Academic A.V. Kalyayev Scientific-Research Institute of Multiprocessor Computing Systems of the Southern Federal University and the Supercomputers and Neurocomputers Research Center are developing high-performance configurable computing systems based on field programmable gate array (FPGA), including computation module "Pleiada," "Taygeta," "Skat-8," and desktop configurable computing blocks "Kaleana" [11, 16, 17].

For example, computer "Skat-8" contains 8 FPGAs of the Virtex UltraScale family with a logical capacity of no less than 100 million equivalent gates each. In addition, the module includes two sections: the first one contains 16 fully submerged in electrically neutral liquid cooling agent of computation module board with power consumption of up to 800 W each, and the second one contains pumping group and heat exchange units that provide channel and chemical cooling. The performance characteristics for the advanced computer "Skat-8" and corresponding configurable computing systems are given in Tables 3.8 and 3.9.

The following system software has been developed in the A.V. Kaliayev Research University of multiprocessor from the Southern Federal University and research and development center of supercomputer and neural computers for effective organization of computation:

- Metacompiler of COLAMO programming language, translating source code, written in this language, into information graph of parallel application

Table 3.8 "Skat – 8" characteristics

№	Characteristics	Value
1	"Skat-8" computer performance	105 TFLOPS
2	Computing rack performance based on the "Scat-8" computer	1 PFLOPS
3	"Scat-8" computer power consumption	13 kW
4	Computing rack power consumption based on the "Scat-8" computer	105 kW

Table 3.9 Configurable computing system characteristics

№	Year of development	Computer name, FPGA family	Performance		
			Board (PI32/ PI64), GFLOPS	Computer (PI32/ PI64), GFLOPS	Cabinet 47 U/ PI64, TFLOPS
1	2009	"Orion-5," Virtex-5	250/85	1000/340	19.2–28.8
2	2010–2012	"Rigel," Virtex-6	400/125	1600/500	34.5–51.8
3	2012–2013	"Taygeta," Virtex-7	900/300	3600/1200	68–100
4	2015–2016	"Skat," UltraScale	7250/2500	82,500/30000	1000–1250

- Sequencer of scalable design at the FPGA logic cells level Firel Constructor, displaying the information graph obtained from metacompiler of COLAMO programming language on the configurable computing systems architecture, placing represented decision on FPGA chips, and automatic synchronization of information graph fragments in different FPGA chips
- IP-core libraries corresponding to COLABO language statements (functionally complete structurally-realized hardware devices) for various subject domains and interfaces for coordinating the information processing rate and network into a common computational structure
- Debugging tools and programs to monitor and access the configurable computing systems state

The structural solutions developed at the abovementioned institute based on Xilinx Virtex UltraScale FPGA and liquid cooling have attained a real performance 1 PFLOPS with the form of standard compute cabinet with 47 U high and power consumption 150 kW. In practical terms, these configurable computing systems can be considered as a basis for developing a national supercomputer with ultrahigh performance.

3.2.8 JSC "NICEVT"

Joint-stock company "Research and development center of electronic computer field" (*JSC NICEVT*) is developing the ultrahigh performance computational platform ES1740.0001 and a high-speed interconnector based on VLSI ES8430 (Fig. 3.7). The JSC "NICEVT" formerly participated in the Union of Russia and Belarus programs "SKIF," "SKIF-GRID," and "TRIADA" as a principal investigator from the Russian side.

JSC "NICEVT" conducts research and develops:

- Data communications equipment based on ES8430 VLSI router for organization of high-speed disruption-tolerant network
- Special software (LUSTRE file system) and optimized program libraries (SHMEM, MPI), standard mathematical libraries (BLAS, LAPACK, SCALAPACK, FFTW, etc.), as well as PGAS-class concurrent languages (UPC, CAF) with support for application program development
- High-performance computational platforms (modification of "Angara" computational platform)
- Functionally complete high-performance multiprocessors

At JSC "NICEVT" the development of "Angara" carrier network with 4D-top topology[18] is ongoing; this can become the basis for creating national technologies relevant for developing a supercomputer. "Angara" uses the powerful microprocessor models (AMD *Bulldozer/Piledriver*) and a communication network of its own design.

3.2.9 KVANT and the M.V. Keldysh Center

Moscow Research University "Kvant"[10] (under the guidance of M.V.Keldysh) together with the M.V. Keldysh Institute of Applied Mathematics developed the RAS[11] computing systems of the MVS family (*MVS-100 and MVS-1000*).

The supercomputer *MVS-1000 M* with massive parallelism is based on dual-processor modules, including processors Alpha 21,264 with a 667 MHz clock frequency and a peak performance of 1.3 GFLOPS. The shared memory of dual-processor module is 2GB. The basic tripod block consists of 64 dual-processor modules, and each of the six base blocks consists of 768 processors. Myrinet local network with a channel traffic of 21,250 Mbit/s in duplex mode is used for interprocessor communication and switching.

In 2010 institutes developed the low-dormant carrier network "MVS-Express," which has a high rate of issuing short messages and is used in the construction of commercially available microprocessor chip-computing systems with distributed shared memory. In contrast to most commercially available

---

[10]http://www.rdi-kvant.ru/Branches.aspx/

[11]http://www.kiam.ru/MVS/research/

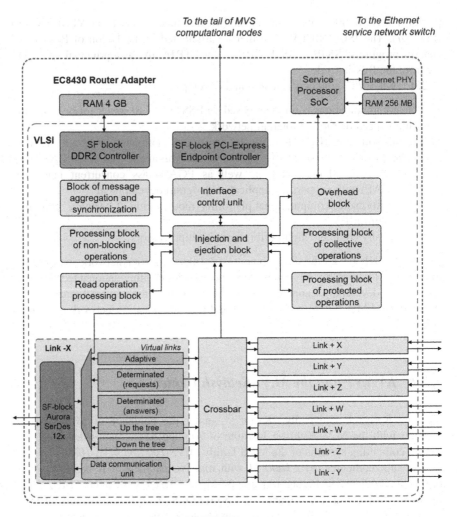

Fig. 3.7 Schematic diagram of "Angara" ES8430 router

networks, "MVS-Express" network is built on the technology of direct switching
of PCI-EXpress packets. As a result, there are no delays associated with network
packets translating in different hardware environments. The network implements
a high-volume distributed shared memory at hardware (Fig. 3.7), which consists
of publicly accessible system memory areas of each of the node cluster and is
fully accessible to each of the nodes.

In 2012 the second "MVS-Express" network generation based on PCI-Express
2.0 was authorized for commissioning. The operations on establishment of a third
network generation based on PCI-Express 3.0 were completed.

The "MVS-Express" network was applied in the construction of K-100 cluster at the Institute of Applied Mathematics named after M.V. Keldysh and also in super-computer software and hardware complex at Peter the Great St. Petersburg Poly-technic University. The active studies are focused on effective shared use of the carrier network "MVS-Express" and InfiniBand to develop advanced architectures for a hierarchical exascale supercomputer.

3.2.10 Program Systems Institute of RAS

Program Systems Institute of RAS was established in 1984 as a branch of the Cybernetics Problems Institute of the USSR Academy of Science (Fig. 3.8.). In

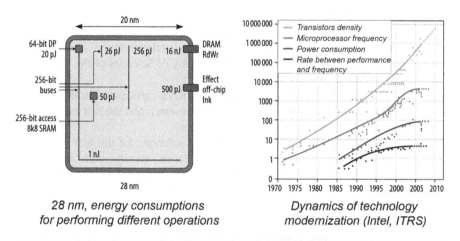

28 nm, energy consumptions for performing different operations

Dynamics of technology modernization (Intel, ITRS)

Design rules	45 nm (2008)	32 nm (2010)	22 nm (2012)	16nm (2014)	11 nm (2016)
Transistor density	1.75				
CPU frequency control	15 %	10 %	8 %	5 %	4 %
Voltage scaling	-10 %	-7.5 %	-5 %	-2.5 %	-1.5 %
Electric capacity dimension	0.75				
Leakage factor	From 1X to 1,43X				

Petascale Computing   Exascale Computing

Fig. 3.8 Future prospects and special aspects of the development microprocessor chips technologies

1986, it received an independent status as the Program Systems Institute of the USSR Academy of Sciences. In the national context of developing supercomputer technologies, the institute has made significant contributions particularly in the field of promising approaches to parallel account organization:

- Programming language Norma – recording the numerical method for problem solving in mathematical physics
- DVM, HDVM – unified complexes for development of scientific and engineering computation of parallel programs
- Programming language T++ and development system *Open TS* – automatic dynamic program multisequencing and hybrid modes support
- Parallel programming language *JNDP* – support for concurrency of embedded (structured) data (under a contract with Huawei)
- "Kentavr" system – automatic programming of account allocation and data communication for grid tasks using cluster systems, including hybrid nodes

The institute's work on supercomputer subject can be broken down into several stages [1, 4, 6–8, 11, 19]:

- 1984–1992: the Program Systems Institute of RAS participated in software development for a multiprocessor with a dynamic architecture *ES 2704* (original national development of FRI ASUSSR, Academy of Sciences of the USSR and NICEVT)
- 1990–1995: the Program Systems Institute of RAS participated in the Russian Trans-computer Association; the beginning of research and the first experiments in the areas that led to the creation of a T-system
- 1994–1998: the search and implementation of solutions for the first version of T-system components; various PC network forms, starting with an original network based on accelerated (up to 1 Mbit/s) RS-232 line and their own communication devices for such connections ending with a cluster based on Fast Ethernet (100 Mbit/s) used as a hardware base
- 1998–1999 years: development of the first version of T-system, establishment of cooperation with Minsk colleagues, and organization of Union state supercomputer program "SKIF"
- 2000–2004: implementation of Union state supercomputer program "SKIF" (Institute of Software Systems RAS has been a general contractor from the Russian Federation) and development of Series 1 (2000–2002) and Series 2 (2003–2006) "SKIF" family supercomputer, software development
- 2005–2007: proactive work in the field of supercomputers and GRID-systems, development of new version of T-system (OpenTS), establishment of Union supercomputer program "SKIF-GRID," and development of scientific and technological reserve for Series 3 supercomputers "SKIF"

Table 3.10 Supercomputers of the SKIF family among Russia's most powerful supercomputers

№	Dates	TFLOPS	Supercomputer	Developer
1	2002/06	0.734	MVS 1000 M	FGUP "Research Institute Kvant," MSC RAS, FGA Federal Research Center of M.V. Keldysh Institute of Applied Mathematics
2	2003/11	0.423	SKIF K-500	SKIF – cooperation
3	2001/11	2.03	SKIF K-1000	SKIF – cooperation
4	2007/06	9.01	SKIF Cyberia	SKIF – cooperation
5	2008/06	12.2	SKIF Ural	SKIF – cooperation
6	2008/06	47.1	SKIF MSU "Chebyshev"	SKIF – cooperation
7	2009/11	21.8	SKIF-Avrora SUSU	SKIF – cooperation
8	2009/11	350	Lomonosov	T-platform Group
9	2012/11	376	Tornado of Joint Supercomputer Center of Russian Academy of Sciences	"RSK" Group
10	2012/11	147	Tornado SUSU	"RSK" Group
11	2014/06	320	Lomonosov-2 (A-class)	T-platform Group
12	2014/11	658	Tornado SPbPU	"RSK" Group
13	2014/11	289	GPU Blade Cluster UNN	Niagara Computers, Supermino
14	2014/11	170	PetaStream SPbPU	"RSK" Group

- 2007–2012: implementation of Union state supercomputer program "SKIF-GRID" (Institute of Software Systems RAS was a general contractor of the Russian Federation) and development of Series 3 and development of implementation approaches for the Series 4 "SKIF" family supercomputers

Nowadays, 75–80% of Russian supercomputers are represented by the "SKIF" supercomputer family ("SKIF K500," "SKIF K-1000," "SKIF Cyberia," "SKIF Ural," SKIF MSU "Chebyshev," and "SKIF-Avrora SUSU"), etc. Four series of "SKIF" family supercomputers are developed (Table 3.10). Thus Series 2 was represented by a supercomputer with a performance from 2.5 to 5 TFLOPS, and Series 3 was represented by a SKIF MSU "Chebyshev" supercomputer with a performance from 60 to 150 TFLOPS.

3.2.11   OOO SPA "Rosta"

OOO SPA "Rosta"[12] is developing a configurable computing system and units based on FPGA Virtex-7 family. For instance, the RB-8 V7 block is made with 1 U

---

[12]http://www.rosta.ru/about/

constructive that provides fixing into the rack. The RB-8 V7 block case has two RC-47 board with radiators, power unit, diagnostic module RDM-03, and air cooling system. Each RC-47 has four FPGAs, connected by a PCI Express bus with the use of a switch. Two switch ports are connected to the cable connectors for scaling and communication with the control host computer.

3.2.12 "T-Platforms," RSK, "Niagara,"and "Immers" Companies

Supercomputers produced by "T-Platforms"[13] have repeatedly been included in the TOP500 rating of the world's top performing supercomputers.[14] In 2015 "Baikal Electronics," "T-Platform's" subsidiary, supported by "Rostekhnologii" and "Rosnano," designed a line of Baikal microprocessors: 8-core Baikal M and Baikal M/S for PC and microprocessors (2 GHz processor clock, 64-bit ARMv8 Cortex A-57 core, 28 nm technological processes) and 16-core according to the 16 nm technology for servers.

The RSK group of companies[15] has developed the "Tornado" supercomputer located at the Peter the Great St. Petersburg Polytechnic University (226th rating position in TOP500 in November, 2016) with a performance of 0,658 PFLOPS. The following developments have been achieved: RSK micro DPC (from 4 to 32 nodes, up to 60 TFlOPS), RSK mini DPC (from 32 to 306 nodes, up to 0.6 TFLOPS), RSK DPC (more than 2 racks with high density, up to 500–600 PFLOPS), and others.

NIAGARA[16] develops supercomputers based on NVIDIA graphic accelerators. FDR InfiniBand network (up to 56 Gbit/s) is used for linked supercomputer computational nodes. The computer can be scaled up to 20,000 computational nodes and contains up to 40,000 Intel Xeon IvyBridge processors and up to 60,000 Nvidia Kepler K40 with a performance of over 90 PFLOPS.

IMMERS[17] develops IMMERS 8 R5 supercomputers with heterogeneous configuration (CPU + GPU) and a peak performance of up to 122.5 TFLOPS. IMMERS supercomputers have high energy efficiency, and energy efficiency coefficient reaches the PUE value 1.05 (in the air system PUE 1.47). The nodes packaging has been innovatively changed to achieve higher density, intercoolers have been added, and also monitoring software and liquid exchange control have been improved.

---

[13]http://www.t-platforms.ru/

[14]Petrenko A. S., Petrenko S. A. Super-productive monitoring centers for security threats. Part 1 // Protection of information. Inside. - 2017. - No. 2. - P. 29-36

[15]http://rskgroup.ru/

[16]http://www.niagara.ru/about/

[17]http://immers.ru/

3.2.13 Lomonosov Moscow State University

The Lomonosov Moscow State University[18] supercomputer complex conducts fundamental research in hydromagnetics, hydro- and aerodynamics, quantum chemistry, seismic, medicine computational modeling, geology and materials sciences, fundamental principles of nanotechnologies, engineering sciences, cryptography, and many other areas.

The computational center of Lomonosov Moscow State University includes the supercomputers "Lomonosov-2" with a performance of 2.102 PFLOPS and "Lomonosov" with a performance of 0.902 PFLOPS, "Chebyshev" with a performance of up to 60 TFLOPS, and IBM Blue-Gene/P with a peak performance of more than 27 TFLOPS. Placing these supercomputers into operation has solutions possible to a number of important tasks for the Russian industry. For instance, for Rocket and Space Corporation Energia, a flow analysis calculation of advanced spacecraft "Rus" during earth aerobraking and landing on its surface were made. For the I.I. Afrikantov Experimental Mechanical Engineering Design Bureau, the problem of mass heat transfer in the separation of sodium oxides device in the first circuit of advanced nuclear reactor has been solved.

The Supercomputer Consortium of Russian universities[19] was organized in December 2008, and today it consolidates more than 50 permanent and associated members including the largest Russian universities. The Consortium has become the main contractor for the project "Supercomputer Education" of the Russian Federation Presidential Commission for the Modernization and Technological Development of the Russian Economy.

According to S.M. Abramov, director of the A.K. Alimazyan Institute of Software Systems, the Russian Academy of Sciences has a significant scientific backlog for the creation of a national exascale supercomputer, which should be developed in the following main areas [1, 11, 19].

Development of computational nodes with advanced architecture. Including the use of FPGAs or dedicated processor (CPU accelerators, FPGA, etc.) for solving problem situation associated with low efficiency of the most known processors and computation accelerators (Fig. 3.9). There is a positive experience of using accelerators and development of corresponding system software and application software in the high-efficiency computing systems SKIF – Avrora (South Ural State University, Chelyabinsk), RVS-KP (Keldysh Institute of Applied Mathematics-KIAM), "MVS-Express," "K-100" (KIAM of RAS with partners), etc.

Creation of national advanced system area networks (in order to replace Infiniband FDR) including realization of hardware support like prospective approach to multiprogramming, for example, one-sided communication, PGAS, and vSMP; various operations during interprocessor communication, synchronization, collective operations, and "calculation in network"; adaptive routing (bypassing failed or

---

[18]http://www.msu.ru/lomonosov/science/computer.html/
[19]hpc-russia.ru/

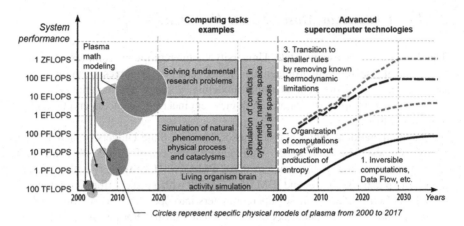

Fig. 3.9 Forecasts of growth in productivity requirements and produced supercomputer's possibilities

overloaded network elements); and routing reconfigurable for the task's specifics. There is a positive experience in the development of the MVS-Express communication networks (KIAM of RAS in cooperation with FGUP "Research Institute Kvant") for the K-100 supercomputer (KIAM of RAS) and the SKIF 3D-top network (Program Systems Institute of RAS) for the SKIF-Aurora supercomputer (SUSU, Chelyabinsk).

Development of data storage system (*DSS*) for systems that consist of hundreds of thousands and millions of computational nodes and also for problem solving:

- Effective parallel data exchange between data storage system and computing machine
- Support follow-on concept of data storage system for extra-high-performance systems

Creation of national infrastructure subsystems including the development of a supplemental network, service network (monitoring and control for super-large devices), power supply, and cooling subsystems. There is a positive experience in the development of the first-world supercomputer based on standard (i86-compatible) processors with liquid computation electronics cooling and an advanced power supply subsystem "SKIF-Avrora" and a service sensor network of SKIF-Servent family for monitoring and hardware management.

Basic and system software development for national supercomputer with ultrahigh performance, including study and production:

- System software (resource manager, queuing network), "light" and "low-noise" operating system for computational nodes, control points at the OS level, OS service subsystems for monitoring and control, reliability control, accident prevention, fault tolerance, stability, and survivability

- Software (low-level) supporting efficient and fault-tolerance hardware utilization: system drivers and libraries, programming languages, and development system of large-scale parallelism
- Programming support tools for ultrahigh performance systems: profilers, verifiers, analysis system and program transformations, integrated development environment, etc.

In a long-term perspective, studies are required in the following areas:

- Advanced optical fiber splice and their effect on architecture
- Implementation of nonconventional approaches to parallel computation: dataflow, reversible computations, etc.

The relevance of the above mentioned technologies is justified by the need to change the physical principle of computational organization in the transition to smaller design standards taking into consideration the thermodynamic constraints:

1. Landauer-John von Neumann's principle: the energy liberation release per operation is no less than $kT\ln2$, where k is the Boltzmann constant and T the absolute temperature (Figs. 3.10 and 3.11).
2. Landauer-Bennett-Merckle principle: in case of reversible operations, energy liberation can be reduced to the level that is necessary for information exchange (with a superconducting element, this can in principle be brought to zero).

The development will be required for computation implementation almost without production of entropy, possible only for reversible computations:

- New technologies and solutions
- New reversible computations basic algebra (the essential scientific reverse has been in A.K. Ailamazyan RAS Institute of software systems)
- Reversible computational logic
- System and languages of reversing programming
- Models and application implementation methods in terms of reversing programming, etc.

In May 2016, I.A. Kalyaev noted at the RAS presidium:

Modern supercomputers consume a lot of energy – about 10 MW. For example, supercomputer Tianhe-2 (holds the second place in TOP500 rating as at November 2016) consumes 17.8 MW! 1GW of power (this is approximately one-sixth of the entire energy output of the Sayano–Shushenskaya hydroelectric power station) and 250,000 cubic meters of equipment (a building with a 50×50 m foundation and 100 m high) is supposedly required for achieving exascale using current technologies. The principal problems with dissipation of heat also haven't been solved.

Additionally, most modern supercomputers function as computing clusters with a quite rigid architecture. As a result, such computation systems demonstrate high performance only when solving loosely coupled tasks, which do not require the exchange of large amounts of data, in other words, those that can be comparatively easy decomposed into mutual unjointed subtasks. When solving strongly coupled

Fig. 3.10 Landauer's limit. The dynamics of reducing the processing costs of a single bit

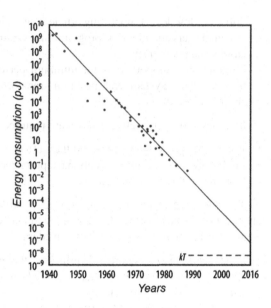

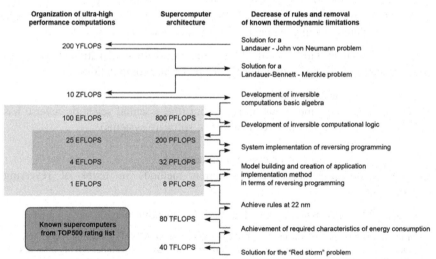

Fig. 3.11 Physical performance limit of nonreversible supercomputers – "Sterling point" (600 kW)

problems (i.e., digital signal and image processing, mathematical physics, symbolic processing, etc.), real supercomputer performance drops sharply and achieves no more than 5–10% of the reported peak performance.

Supercomputer performance decrease is also observed when the number of processors in the system increases. In the rigid architecture of computing clusters, there are large unproductive time costs associated not with useful calculations but rather with the organization of the computing process: some of the processors are idle, and some are engaged in the information transit. For example, modeling and optimizing the operating modes of a gas turbine engine will require 2500 days of machine time, which is equivalent to almost 7 years of using a supercomputer with a performance of 1 PFLOPS (equivalent to a second Russian supercomputer "Lomonosov" with 0.902 PFLOPS)!

I.A. Kalyaev proposed solving this problem by developing application-oriented computing processor systems 128, 143, 172]], which can combine the advantage of the cluster's multiprocessor, taking into account their universality, with specialized ones, which have a high real performance. This would allow the user to design the computing system's architecture. "The reconfiguration ideas were used earlier in incremental computers, homogeneous computing environments, multiprocessors with programmable architecture. However, it never really took off due to a lack of hardware components that correspond to the concept of reconfigurable architecture," he noted. "Such hardware components appeared at the beginning of the twenty-first century, the so-called FPGAs with high-scale integration."

Owing to this innovative approach, it has now become possible to significantly boost the efficiency of national computation systems:

- Real peak performance rate by 5–10 times
- Specific performance (i.e., unit of volume performance) by 100–150 times
- Performance energy efficiency per watt of power consumption by 5–10 times
- Fault tolerance (i.e., gamma-percentage of error-free running time) by 1.5–3 times

3.3 Creating the Computer of the Future

In the national scientific school of information security, the necessity arose to design a cognitive high-performance supercomputer in early 2015. In considering research possibilities, studies began exploring new principles and a possible supercomputer architecture based on "computational cognitivism," in which cognition and cognitive processes are forms of symbolic computation. As a result, a general functional model of the cognitive supercomputer was proposed in addition to several private prototype models of hardware-software complexes of Monitoring in the Detection,

Table 3.11 Differences between existing computer and "computer of the future"

№	Classic computing system	Cognitive supercomputers
1	Accurate algorithm description is required (symbolic computing)	Data processing algorithms are similar to signal processing algorithms; instead of a program, it has a set of neuron weights, instead of programming – training of neurons (weight setting)
2	Computer data must be correct. Computing system has insufficient stability and survivability under information confrontation conditions. Computing system memory goes into denial of service in case of destructive effects on it	Cognitive system is resistant to impacts and various noise; data corruption has no significant effect on the result (including multiple neuron failure)
3	Each work object is explicitly specified in the computer's memory	Work objects are unobviously represented by neuron weights. As a result, the cognitive system can work with objects that it has not met before and generalize training results
4	Pattern recognition and associative information search algorithms are characterized by high labor coefficient and computational complexity	Pattern recognition and associative information search algorithms are optimal according to labor coefficient, reliability, and complexity

Prevention and Response to Cyber Attacks (SOPCA). The solutions obtained differ notably from well-known solutions based on "Von Neumann architecture" in their unique ability to independently associate and synthesize new knowledge about quantitative and qualitative information confrontation factors [3, 4, 6, 15, 20].

3.3.1   Relevance of the Problem

Currently, the worldwide number of digital data is increasing by at least 60–80% annually. At the same time, the computational capabilities of the known evolutionary "Von Neumann architecture" modifications have reached their maximum capacities. For this reason, technologically advanced countries are actively conducting research in the field of "computer of the future," based on the principle of the organization of a living brain, capable of long-term learning in the real world (Table 3.11). This will give rise to a conceptually new class of *artificial cognitive systems*, a new computational architecture paradigm with various applications for all spheres of human activity. Similarly, new industrial fields will appear. According to IBM projections, such a "computer of the future" with sufficient computational resources for modeling the human brain, should appear by 2019–2020 (Figs. 3.12 and 3.13).

With this goal in mind, in March 2013, the European Commission allocated a "technology of the future" grant in the amount of 1.3 billion euros for the "Human Brain Project," an international scientific project designed for 10 years. In April of the same year, Barack Obama announced the beginning of a national science and

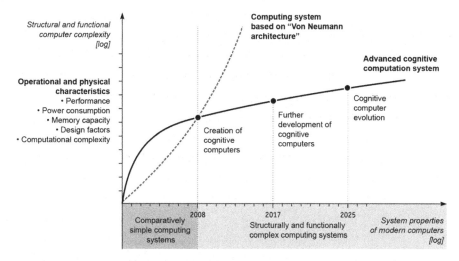

Fig. 3.12   A new era of cognitive computations, IBM

technology initiative called *BRAIN Initiative* with the aim to research the human brain structure with 3 billion dollars[20] in funding.

The US research program *Machine Intelligence from Cortical Networks* (*MICrONS*) stands out among the well-known projects on the given topic (Fig. 3.14). This program aims to model a living organism's brain activity and, most importantly, develop prototype models of a cortical computer, which would be organized and operated as a brain cortex. The *MICrONS* program is oriented at developing universal cognition in computers, engineered for creating international cognitive computations, in contrast to the highly specialized previous emphases of object recognition and classification (Fig. 3.15).

Nowadays, more than a quarter of DARPA research programs and more than half of the US IARPA programs are dedicated to solving the scientific-technical problem of creating a *special purpose artificial cognitive system*. On July 8, 2013, IARPA published a request to create the connectome modeling technology as well as cortical column functions as an initial module of the brain of a living organism. The SN-13-46 contest *Request for Information (RFI) on Research and Development of a Cortical Processor* was announced by DARPA on August 14, 2013. The main goal of this project is to create an artificial cerebral cortex of a living organism.

At the international level, the Chinese government launched the research program "Brain Understanding" in 2014–2015 with 11 billion dollars in funding over 10 years. Dozens of Russian scientific institutions are undertaking research in this field, and many public-private partnerships have arisen with a view to creating an artificial cognitive system.

---

[20]http://www.neuroscience.ru/showthread.php?t=5669/

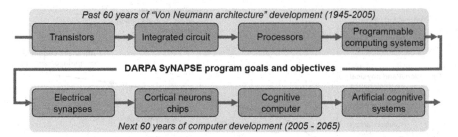

Fig. 3.13 Expected results of cognitive technological development, IBM

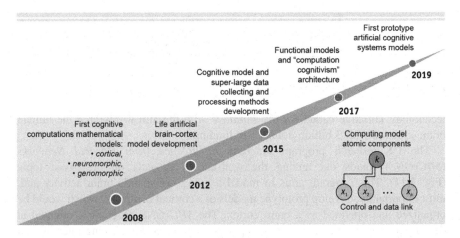

Fig. 3.14 MICrONS program roadmap, USA

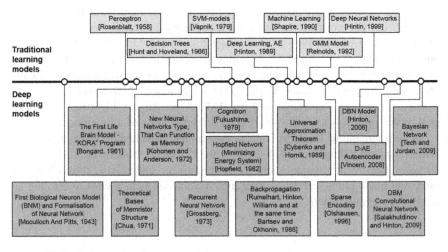

Fig. 3.15 Evolution of deep learning models

3.3.2 Existing Reserve

In his famous 1984 monograph [21], N. Wiener proposed the mathematical models for investigating properties of data reduction processes in living organisms. Ten years later, F. Rosenblatt invented a single-layer perceptron for solving the classification problem and then created the first *Mark-1* neurocomputer (1960) [22–24]. In the early 1960s, mathematician R. Block formulated the *recognition theorem*, and radio engineer B. Widrow developed the first *ADALIN* artificial neural networks (adaptive adder), which is now the standard component of each signal processing system [25, 26].

The "Kora" program is one of the first living brain models, developed for the M-2 computer [23] in the USSR under the guidance of M. M. Bongard in 1961. However, the functionality of the first living brain software and hardware models were quite minimalistic due to the limited capacity of the microelectronics hardware components in the 1960–1970s. As such, neural networks theory did not bring revolutionary changes in adaptation and optimal control algorithms at that time (Fig. 3.16).

The theoretical bases of memristor (memory and resistor) construction were developed by the professor of physics L. O. Chua from the University of California, Berkeley in 1971. The memristor became the fourth fundamental electronics element along with *resistors, capacitors, and inductors*. Here, a "memristor" is a two-terminal component, whose electrical resistance reversibly changes from a high-resistance state (HRS) to a low-resistance state (LRS) depending on the current flowing through it. The connection is formed in the neural network nodes by reducing the memristor resistance, the stability and weight of which depend on the memristor properties. Forgetfulness and slowing down take place with an increase in memristor resistance. In this device, information and control signals can be integrated, and the system is also capable of self-learning. Its nonvolatility property

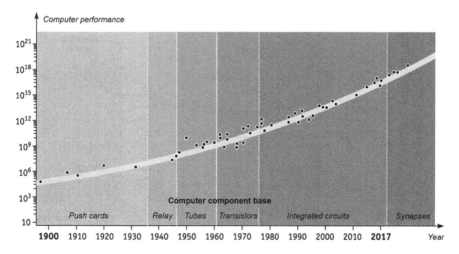

Fig. 3.16 Computer aids development evolution

constitutes a significant difference (and an advantage at the same time) between memristors and most known types of semiconductor memory. At present, two fields of application are found for memristors: nonvolatile general-purpose memory, designed to replace existing RAM and ROM, and solid neuromorphic electronics, development of devices, in which the passage and change of data signal follow the same logic as the processes in a living nervous system. Cortical structures, emulating the work of brain cortical columns, can be considered as the highest point of development of the latter area. Essentially, memristors can be used not only as synapses but also as an element for adding artificial neurons.

In 2008, the first memristor prototype model was based on titanium dioxide (TiO2), a semiconductor, having a high resistance in pure form, and was created under the supervision of Stanley Williams in the Hewlett-Packard Research Laboratory. In April 2010, Hewlett-Packard managed to create memristors with a side of 3 nm, and switching speed approximately equal to 1 ns. 3D block of memristors makes it possible to place 20GB of data in a volume of 1cm^3.

Hewlett-Packard and Hynix companies (HLR) consortium initiated the serial production of storage memory technology based on memristors. Furthermore, HLR continued to improve the neuromorphic chip by making it possible for neurons to independently change the impulse frequency (as the living brain does), and IBM, using data on the brain cortex structure, developed a new method of linking artificial neurons to each other and got closer to the brain of a living organism, in which 1 cubic centimeter contains about 10 billion synaptic connections.

Figure 3.17 provides an illustration of a potential artificial neuron model, consisting of dendrite (branch that extends away from nerve cell body and represents neuron input connections) and synapse (adder inputs) analogues, as well as its own nerve cell body analogue (adder) and axon analogue (long apophysis that extends away from one of the sides of soma and serves to transmit an output signal to other neurons, which are adder outputs).

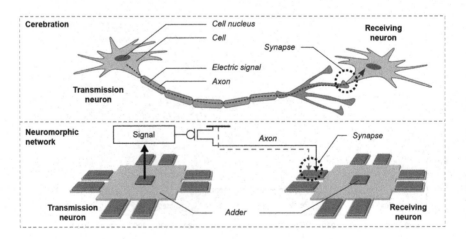

Fig. 3.17 Projected model of a living organism's brain

Nowadays, the main memristor types are as follows:

- Solid-state inorganic semiconductor memristors based on metal and silicon oxide, for example, Al: TiO_2 ($Ti_{0.85}Al_{0.15}O_y$ or $Ti_{0.92}Al_{0.08}O_{1.96}$) with 20 nm thickness, in which the link efficiency change occurs due to a doping agent drift in the electric field
- Memristors based on organic material, in which the link efficiency change is provided by an oxidation-reduction reaction

As can be seen above, the first memristors became the first nonliving analogue of living connectivity – synapses – and then became elements for adding artificial neurons.

Inorganic memristors were obtained on "Nanofab-100" NT-MDT equipment (under the guidance of V.A. Bykov) at Tyumen State University (titanium dioxide, April 2012), the Russian Research Centre Kurchatov Institute and in the F.V. Lukin Research Institute of Physical Problems. The first samples of organic polymeric and composite memristors were obtained at the A.N. Nesmeyanov Institute of Organoelement compounds Russian Academy of Sciences and National Research University of Electronic Technology.

A timeline of computer technology from neurocomputers to cognitive computers is shown in the Inset 3.2.

Inset 3.2 From neurocomputers to cognitive computers

1943 – W. McCulloch and W. Pitts formalized the neural network terms and proposed the first biological neuron model (BNM).

1948 – N.Wiener published his famous "Cybernetics" monograph.

1949 – D. Hebb proposed the first learning algorithm.

1958 – F. Rosenblatt invented a single-layer perception for solving the image recognition problem.

1960 – B. Widrow in cooperation with M. Hoff developed ADALIN for predicting and solving the adaptable control problem.

1961 – "Kora" program, one of the first models of a living brain, was developed for the M-2 computer under the supervision of M. M. Bongard.

1963 – A. P. Petrov carried out research on the "difficult" tasks for the perceptron at the Institute for Information Transmission Problems of the USSR Academy of Sciences.

1969 – M. Minsky published a formal proof of perceptron limitation and showed that a perceptron cannot solve some problems ("parity" and "one in the block" problems) associated with the representation invariance.

1971 – The theoretical base of memristor construction was developed by professor of Physics L. O. Chua from University of California, Berkeley.

1972 – T. Kohonen and J. Anderson independently proposed a new type of neural networks that can function as memory.

(continued)

Inset 3.2 (continued)

1973 – B.V. Khakimov proposed a nonlinear model with synapses based on splines and implemented it for solving problems in medicine, geology, and ecology.

1974 – P.J. Werbos and A.I. Galushkin simultaneously invented a backpropagation for multilayer perceptrons.

1975 – Fukushima presented a cognitron, a self-organizing network, designed for invariant image recognition (by remembering virtually all image states).

1982 – J. Hopfield demonstrated that a recurrent neural network can be a minimizing energy system ("Hopfield network"). Kohonen presented a network system, which can learn without a teacher (Kohonen neural network), solving data clusterization and visualization problems (self-organizing map) and other problems of initial data analysis.

1986 – Backpropagation was rediscovered and essentially expanded by D.E. Rumelhart, G.E. Hinton, and R.J. Williams independently and simultaneously with S.I. Bartsev and V.A. Okhonin (Krasnoyarsk group).

1988 – The first i80170NX analog neuroprocessor prototype model was developed by Intel. One year later, Intel (in cooperation with Nestor and US DARPA financial support) started developing NI1000 digital neurochip that was announced as i80160NC.

1998 – The Russian company NTC "Module" that created the NM6403 neuroprocessor entered the international neurochip market. Designed by Russian engineers, this neuroprocessor is produced at Samsung.

2000 – Various highly parallel neural accelerators (coprocessors) were widely used. The solutions by Motorola, Echelon, IBM, Siemens, Fujitsu, and others were represented at the neuroprocessor market. At the same time, there were no universal neurocomputers; they are created for specific tasks.

2007 – J. Hinton created an algorithm for deep training of multilayer neural networks at the University of Toronto. Hinton used a Restricted Boltzmann Machine (RBM) for training the lower network layers. After training, it became possible to use a fast ready application, capable of solving a specific problem (e.g., face scan on images).

2008 – Under the supervision of S. Williams, engineers from the Hewlett-Packard development laboratory created a prototype model of a memristor, an electronic component based on titanium dioxide, which is an artificial analogue of connectivity or "synapses," which can store and simultaneously process data as a living brain does.

2011 – A new architecture and a number of subsystems of the "cognitive computer," capable of independent learning, was introduced by IBM, Hewlett-Packard, and HRL (Howard Hughes) laboratories under the US DARPA Program: SyNAPSE.

(continued)

Inset 3.2 (continued)

2012 – Google researchers launched a neural network on a 1000-server cluster (16,000 processor cores, 1.7 billion synapses) to improve the accuracy of speech recognition by 20–25%.

2012 – Computer technology developers achieved the technological hardware component density necessary for modeling the human brain.

August 2, 2013 – RIKEN (www.riken.jp) scientists simulated the 1-second work of 1% of the human brain, a network of 1.73 billion neuronal cells and 10.4 trillion synapses. For this purpose, 82,944 processors and 1 Pbyte (one thousand terabytes) of K-computer memory, a well-known Japanese super-computer produced by Fujitsu, were used. The computations took 40 min.

2013 – NVIDIA created the largest artificial neural network based on its own GPU in cooperation with the team from Stanford University. The mentioned neural network was 6.5 times larger than Google's neural network. To create it, 16 servers based on NVIDIA graphics processors (11.2 billion parameters) were used. It was created to study the human brain learning process.

2016 – Russian JSFC "Sistema" (OJSC "RTI" in partnership with the Information Security Center of Innopolis University) developed the first proto-type model of an open segment of the national early-warning system for cyber-attacks. The solutions obtained differ notably from existing SOPCA (SPOCA) solutions in their unique ability to independently associate and synthesize new knowledge about quantitative and qualitative information confrontation mechanisms. Particularly, it made it possible to synthesize early warning and opponent deterrence scenarios in the cyberspace of Russian Federation on extremely large volumes of structured and unstructured information from a variety of Internet/Intranet and IIoT/IoT sources.

3.3.3 IBM Deep QA "Watson"

Initially, the IBM Deep Question-Answering (Deep QA) "Watson" (henceforth referred to as Watson), named after the company's founder, Thomas Watson, presented an expert system (2006), containing a large corpus in a natural language, a system of various complex queries to place, and algorithms for choosing the best possible response there to. Watson's key feature is the ability to process not only structured but also unstructured data, which currently accounts for about 80% of all global information resources.

Today, Watson is a hardware-software complex based on Apache UIMA source environment and a cluster of 90 IBM Power750 servers, each equipped with 8-core processors of POWER7 architecture with a total RAM of more than 15 terabytes. Watson is designed for "deep" natural language processing (in oral and written form)

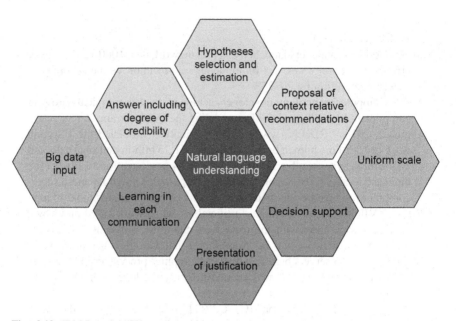

Fig. 3.18 IBM DeepQA "Watson" working process

taking into account external parameters including the emotional state of the human asking questions. The input is a text, and the output is enriched XML files. After receiving a question, Watson implements a syntax and semantic question analysis in natural language (NLP), generates a series of proposals and hypotheses (KP&R), processes data from additional data resources (data and knowledge base) as necessary, searches and ranks possible answers (IR), implements machine learning (ML), and gives a justified answer to the question. Today, Watson includes more than 40 key programs written in Java, C++, and Prolog (using Apache Hadoop) (Fig. 3.18).

It should be noted that Watson is not a neural network in pure form, but it implements some information processing principles of living cortical columns, which were tested as part of IBM research studies under the Blue Brain Project. For example, answer correctness is evaluated by taking into account the total numbers of words between the question and sentence hypothesis. Additionally, the largest common sequence of words length is calculated, taxonomic object analysis is performed, temporal and spatial data are compared, and an associative information search is performed. Text representation graphs are used where the graph nodes are words and the graph edges express the grammatical and semantic relations between them. Watson gives an estimate of the probability of correctness for each of the possible answers based on a weight coefficient, calculated with system training, after receiving a number of hypotheses. The highest-scored answer variant is given to the user on the condition that it exceeds an expected accuracy threshold. The texts to be analyzed by Watson are first given by users, and after that the system can add new

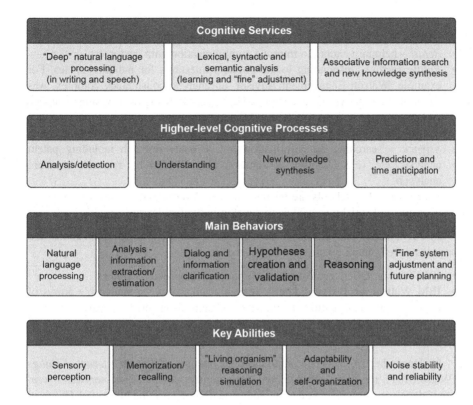

Fig. 3.19   General characteristics of IBM DeepQA Watson

documents from the structured Internet resources (such as *dbPedia, 7WordNet, Yago*), if they contain new, relevant information useful for a correct answer to the given question (Fig. 3.19).

The following priority areas of system development and use in practice are presented on the Watson website:

- *National security* – big data collection and analysis of a wide variety of antiterrorism problem, cyber-attack detection and prevention, fraud investigation, etc. (e.g., for the NSA and FBI national security department).
- *Healthcare and medical insurance* – consulting in the medical area (diagnosis and treatment options) and risk estimation for various departments working in these fields, as well as usage of the *Hippocrates* application from MD Buyline, *virtual concierge service* Welltok, etc.
- *Finance* – helping in investment planning, for example, for Citigroup by a distant cloud computing service engineering implementation (Watson analyzes current customer needs, processes financial and economic information from various sources, and analyzes data given by customers, which results in bringing the digital banking and financial transactions to a completely different level).

- *E-commerce* – choice of recommendations that better match the customer's queries (e.g., Fluid Expert Personal Shopper third-party application).
- *Retail* – helping vendors to improve communication with customers.
- *Public sector* – responding to the public's questions for public agencies. For example, Watson Engagement Advisor, an analytic consultancy complex, was developed for the American company USAA, specializing in providing insurance and consulting services to US service members, who retired from their service and are transitioning to the nonmilitary life, as well as for their family members.
- *Science* – searching for information to accelerate research including patent analysis, etc. (e.g., Johnson & Johnson uses Watson for research literature analysis).
- *Education* – US schools are testing Teacher Advisor with Watson to improve curricula and syllabi, as well as to develop individual learning plans.

In 2014, IBM Watson Group launched new cloud cognitive applications and services development oriented toward corporate users:

- IBM Watson Discovery Advisor to support pharmaceutical studies
- IBM Watson Analysis to identify semantic links between analyzable arrays of large data and presentation of collection and analysis results in a clear visual form
- IBM Watson Explorer to standardize the cognitive software applications development

On December 16, 2015 IBM announced the opening of *Watson Internet of Things* (Watson IoT), the global headquarters for cognitive solutions and services operational development with regard to the Industrial Internet and the Internet of Things in Munich (Germany). The company also announced the opening of eight Watson IoT client centers in Asia, Europe, and the Americas. IBM client centers are located in Beijing (China), Böblingen (Germany), São Paulo (Brazil), Seoul (South Korea), Tokyo (Japan), Massachusetts (USA), North Carolina (USA), and Texas (USA). A corresponding Watson IoT Cloud Platform was presented.

On December 2, 2016, IBM also announced a set of cognitive solutions for marketing (*IBM Watson Content Hub*), commerce (*IBM Watson Customer Experience Analytics and IBM Watson Order Optimizer*), logistics (*IBM Watson Supply Chain Insights*), workforce management (*IBM Watson Workspace*), and others.

3.3.4 Basic Concepts and Definitions

The term *cognitive computer* was introduced in 2008 by IBM research center members who worked on the *Cognitive Computing via Synaptronics and Supercomputing* (*C2S2*) DARPA project. The project primarily aimed to create the first "computer of the future," a prototype model based on the principles of the organization of a living brain [3, 4, 15, 23, 27, 28].

Usually, *artificial cognitive system* refers to a mathematical model of a living brain, as well as its software and hardware implementation with its design based on the principles of a living organism's brain. For example, this can include computing systems with unique abilities of self-learning and new knowledge synthesis by associative recombination of the received data. In particular, this can also include solving regression and classification problems, associative information search and clusterization, understanding the meaning and contextual significance of data, self-learning and analysis of knowledge from various sources, informational support for decision-making, synthesis of hypotheses, modeling of reasoning, etc. It is worth noting that this definition corresponds to the definition from FP7 (7th Framework Programmes for Research and Technological Development), according to which artificial cognitive systems are "human-created systems capable of interactive behavior based on real world adequate perception."

By "cognitive computer," we mean a supercomputer with high (petascale) and ultrahigh (exascale) performance and massive parallelization of operations with the natural ability to store the data measured and make it accessible for later use. The abovementioned "computer of the future" is similar to a living brain in two ways: (1) it acquires knowledge via a learning process, and (2) it uses intensity connectivity values, called synaptic weights, for information storage. Other definitions of a cognitive supercomputer also exist (Table 3.12) and provide a rich semantic range of this definition.

In a cognitive computer, the artificial neurons are modified as a result of learning and store data as synthesized and recoverable variable postsynaptic membrane receptor models. At the same time, these neurons function as data processors when accessed through individual afferent links. The following basic functions are provided: *data record* (receptor synthesis), *data activation* (postsynaptic potential axon value achievement – "threshold excess"), *system data reconsolidation* (new links added to the inputs with the inclusion of molecular or external permitters), *data cancellation* (through incorrect slowing down and gradual receptors disintegration), filtration of essential (by trace migration), etc. (Fig. 3.20). It is estimated that the cognitive computer that will provide a high processing rate on extremely big volumes of structured and unstructured data from various Internet/Intranet and IIoT/IoT resources (Big Data and Big Data Analytics theme).

On August 18, 2011, as presentation of the first phase results of DARPA SyNAPSE (*Systems of Neuromorphic Adaptive Plastic Scalable Electronics*), IBM employees demonstrated a non-programmable chip capable of learning and adjusting the link modifications between neurons. In total, two working prototypes were presented, and each of them was developed according to 45 nm SOI-CMOS design rules and consisted of 256 neurons (one of prototypes – 65,536 self-learning synapses). Therefore, the first significant academic and practical results were obtained in 2011, thus warranting talk of a new era (the third) in the development of modern computer technology (after the tabulating machine era and "von Neumann architecture" evolution modification era).

Currently, more than 400 engineers from IBM Almaden Research Center in San Jose (not far from Silicon Valley) under the supervision of Dharmendra Modha are

Table 3.12 Cognitive computer definition semantics

№	Research area	Neurocomputing system definition
1	Mathematical statistics	A cognitive computer is a computer system that automatically generates a description of random process characteristics and their aggregate or that has complex and often a priori unknown distribution functions
2	Mathematical logic	A cognitive computer is a computer system having an algorithm represented by a logical particular type elements network – neurons with complete or partial renunciation of Boolean elements (i.e., AND, OR, NOT type)
3	Threshold logic	A cognitive computer is a computer system with a problem algorithm represented by a threshold element network with dynamically reconfigurable coefficients and settings algorithms which are independent of threshold element network dimension and input space
4	Computer science	A cognitive computer is a MSIMD – an architecture computer system, in which a uniform structured compute-processing element is simplified to the level of neurons; the links between elements are significantly complicated, and the programming is moved to linkage between the changes in the computational elements weight coefficients
5	Medicine (neurobiological approach)	A cognitive computer is a computer system represented as a model of the interactions in the cell nucleus, axons, and dendrites linked by synaptic connections (synapses) (i.e., model of biochemical processes occurring in the nerve tissues)
6	Economics and financial industry	A cognitive computer is a system providing parallel processing of smart "business transactions" with training elements and generating new knowledge useful for business operations

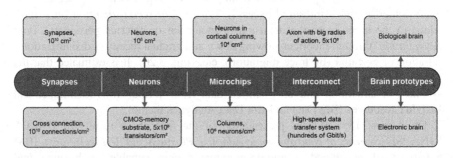

Fig. 3.20 Component base of the cognitive computer

developing cortical (similar to brain cortex organization) neurochips based on the developments of modern-day neurophysiology and the microelectronic. "All in its own good time," said Dharmendra Modha said. "The modern supercomputer hardware components and advances in neurophysiology are what research engineers in the 20th century didn't have."

In 2012, Modha demonstrated a possible implementation of the microcolumn structure – 256 neurons and a memory unit that set properties of 262,000 possible synaptic connections between artificial neurons and placed on 2–3 mm silicon pieces. These artificial neurons and synapses groups were able to process information and respond to input signals similar to living organisms. Moreover, he ran tests of more complex chips, presenting a grid of more than 1 million neurosynaptic cores connected together in a kind of brain cortex.

In the summer of 2015, IBM announced a new software architecture based on modular code blocks, *corelets*, the main idea of which was to release programmers from routine coding for neurosynaptic network. Additionally, they proposed an already functioning environment for programming automation – a design tool that made it possible to select and configure ready-made corelets for different neural network nodes. For example, more than 150 corelets have been developed for various tasks, raning from the recognition of persons to musical compositions in video streams and recordings.

Scientific work has already begun researching chips which can simulate the brain's work under the guidance of Narayan Shrinivas at HLR – a General Motors and Boeing collaborative laboratory in Malibu. For example, a prototype model of the living brain was developed based on 576 artificial neurons that can change synaptic connections depending on processed data (i.e., capable of learning and adaptation, including additional learning in changing real-world conditions).

Most modern cloud services are technically implemented based on classical "von Neumann architecture" computers and at the present time have reached their maximum functional capabilities. This new cognitive computer architecture is required to create and develop principally new adaptive feature and self-organization of these services and further develop functionality (e.g., Google's self-driving car and Apple's virtual assistant Siri). At the same time, using supercomputers with high performance for simulating artificial neural and cortical networks and solving "deep-learning" problems consumes unacceptable levels of power (up to 20 MW and more). Meanwhile, the obtained memristor prototypes, neuromorphic, and cortical computation modules and microprocessors demonstrate the possibility to reduce power consumption required by several orders of magnitude.

Consequently, today, a "cognitive computer" algorithmic basis provides theories of computational systems which are *cortical* (similar to human brain cortex organization), *neuromorphic* (similar to a real nervous system), and *genomorphic* (similar to life genetic and epigenetic mechanisms). Moreover, modern cognitive computational systems can be technically implemented on most existing computational platforms as:

- Artificial neural networks in a software environment based on supercomputers, including as "cloud" systems in global and corporate networks and intranets
- Autonomous technical devices and robots
- Large technological complex control systems
- Neuromorphic electronics devices based on learning nanomaterials and hybrid associative computing systems (as it was technically realized in such projects as

IBM DeepQA Watson, Siri of Apple Corporation, Google's neural network artificial intelligence project and its robots, Japanese JST project, Canadian Spaun project, etc.)

The solution to the mathematical problem in the cognitive basis lies in abstract cognitive mathematical theorems. The main stages of solving nearly all problems in the cognitive basis are input, output, and desired output signal, forming error signals and optimization functionality, forming a suitable cognitive system or network structure, developing an algorithm equivalent to problem solving in the cognitive basis for configuring the cognitive system settings, and conducting research on the decision-making problem solving process. The abovementioned aspects make development of modern control systems with a cognitive approach on the cognitive basis one of the most promising areas in implementing multichannel and multiply connected control systems. In particular, it includes a cyber-attack early-warning system for critically important infrastructure of the Russian Federation based on the collection and analysis of extremely big volume of structured and unstructured data from various Internet/Intranet and IIoT/IoT resources (Big Data and Big Data Analytics theme) [4, 7, 27–29].

3.3.5 Russian Experience

In our opinion, the research group from the University of Tyumen (UT) under the supervision of S.U. Udovichenko and V.A. Filippov,[21] as well as the research group from Russian JSFC "Sistema" (OJSC "RTI" in partnership with the Information Security Center of Innopolis University) under supervision of S.A.Petrenko[22] have achieved quite interesting results.

Researchers from UT together with TASO (Tyumen Associative System Unification) collaborate in the following areas.

1. *Development of a software model for an artificial brain cortex, capable of processing associative base types existing in signal systems and solving various cognitive problems: from answers and questions to synthesizing new knowledge.* The software package of the integrated environment for developing artificial cognitive systems IDEAI is being developed. The software package is implemented in a parallel version based on the client-server architecture for supercomputer clustering architectures. It provides:

- Creation and operational support for artificial biomorphic neural networks with a number of elements (somas, dendrites, axons) more than 200 million units per

---

[21]https://vestnik.utmn.ru/energy/spisok-avtorov/112622/

[22]http://www.mathnet.ru/php/archive.phtml?wshow=paper&jrnid=trspy&paperid=787&option_lang=/

standard eight-core node based on a 3.0 GHz Intel Xeon E5450 (Harpertown) processor type or similar

- A rate of processing data neural network of 5 mln artificial synapses per second at 1GHz of available processor speed
- Fault-tolerant, completely parallel design environment for neural networks and complex systems with distributed multiprocessor system realization on T-Blade type cluster based on MPI and hybrid system, integrating MPI and OpenMP technologies
- The ability to scale software capacity by increasing the number of nodes in the cluster

2. *Creation of a basic software-implemented artificial cognitive system,* in which the artificial cortex is supplemented by functional software models of the retina, lateral geniculate body, peristriate cortex (visual system), auditory subsystem, model of entorhinal cortex, subciclum, hippocampus and dentate gyrus (artificial personality space-time continuum subsystem formation), motive learning subsystem, essence filtration system, active and forced attention module, machine subjectivity module, and other required subsystems.

3. *Designing neuromorphic electronic devices based on solid-state nonorganic semiconductor memristor learning nanomaterials* (memristors based on titanium dioxide thin films were obtained in UT in 2012). The UT employees' design models and methods form the basis for the research – particularly biomorphic neuron models with multiphase memory trace consolidation, cortical column models, cortical (similar to the brain cortex) neural networks, and cybergenomic technology of managing growth and development of super-large artificial neural networks (Fig. 3.21). The "TASO-Neuroconstructor" software package was developed to create neural networks make it possible to:

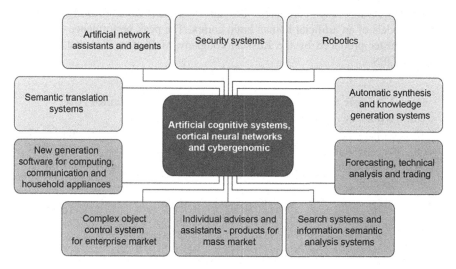

Fig. 3.21 Practical applications of UT cognitive systems

- Build networks with the maximum number of neurons possible on each cluster node (at least hundreds of millions of synapses per node).
- Provide high cortical computation speed, as processing a simple sentence from input to output is thousands of synapses by the shortest path (the packet processing rate must be at least several million synapses per second at 1GHz of available processor speed).
- Create hundreds of unspecified biomorphic neuron model types, including multiphase tracing (from ion transport to structural transformation).
- Provide growth processes and development of the created neural systems (as it is impossible to manually create a network of billions of neurons, even by copying, due to their irregular structure).
- Provide effective input-output interfaces for the basic types of sensory information and motor reactions (i.e., constructing visual fields, sound input, telemetry, movement output, etc. N.B., industry requirements in many cases are measured in microseconds).
- Provide cloud access for researchers and developers to supercomputer resources (e.g., to enter industry-specific standards for representation of neurobiological connectome research results, as well as for technical developments).

"TASO Neuroconstructor" software and its successor IDEAI (Parallel IDE AI Neuro Standard, Neuro Sensorics, Neuro Genomics, Nero Cloud versions, etc.) characteristics are available on the software website.[23]

To solve these problems, UT supercomputers, including "Mendeleev" (11 TFLOPS) by "T-Platform" (Inset 3.3), are used along with the technological platform "Nanofab-100" Module Technological Platform for developing nanotechnological facilities designed under the supervision of V.A.Bykov (Zelenograd) for nanoelectronics production and other university equipment. On December 17, 2013, in Moscow, a new supercomputer "T-Nano" was launched for storing software-based models of an artificial human brain cortex. The peak performance of the "T-Nano" cluster produced by the Russian company "T-Platform" at that time was 220 TFLOPS.

Inset 3.3 Basic Characteristics of the Supercomputer Named After D.I. Mendeleev
The company "T-Platform" proposed the creation of a supercomputer based on the reputable T-Blade 1.1 blade-servers series for "TASO-Neuroconstructor" software package and neural networks developed on its basis use.

(continued)

---

[23]http://www.taso.pro/?page_id=416/

Inset 3.3 (continued)

The supercomputer named after the great Russian scientist D.I. Mendeleev was delivered and set up in 2010. This computing system has 11.5 TFLOPS performance and built on 164 Intel Xeon processors with a 2.9GHz processor speed. The communication network is built on a QDR InfiniBand basis. During project implementation, "T-Platform" carried out a full operating cycle from designing all the computing system components to its commission. Construction, starting-up, and adjustment works were conducted on the computing systems and complex engineering infrastructure of the supercomputer center with tight deadlines. Particular attention was paid to the integration into the system and customization of the client's applied software. At the end of the project, the company's specialists conducted training for administrators and users to show them the settings and system operation. "T-Platform" provides warranty and post-warranty service 24/7.

The hardware-software complex is a high-performance solution based on the T-Blade 1.1 platform and the TB2-XN system, integrating 64 processors (384 cores) in a 7 U high chassis, which is a leader in computational density among systems based on x86 architecture. The high level of components integration on the TB2-XN mainboard, which generates about 570 W heat, requires efficient cooling. Computer analysis yielded the model for optimal radiator design with the best mass-efficiency ratio. The chosen design made it possible to reduce the chassis weight to 153 kg, thus reducing the load on the installation floor in supercomputer centers and optimizing infrastructure setup costs.

The TB2-XN computing system is a universal solution for the highest performance range of supercomputers and is compatible with an extremely broad spectrum of HPC-applications and peta-level scalability. Supercomputers based on TB2-XN are used in solving a vast range of problems: from energy engineering and heavy equipment industry to nano research and drug design, where decreasing the time and development costs are a critical factor of competitiveness. All cluster computational nodes are integrated in high-efficiency QDR InfiniBand system area network with 40Gbit/s of bandwidth and a pilot network Gigabit Ethernet standard. PAC contents include Clustrx monitoring and control system by "T-Platform" for computing system and engineering infrastructure control. The user-friendly graphical interface provides powerful capabilities for monitoring the work of all applications and the processes of computing system elements, as well as load control on the security features.

Automatic emergency outage provides correct system shutdown in case of power supply failure or critical changes of temperature conditions.

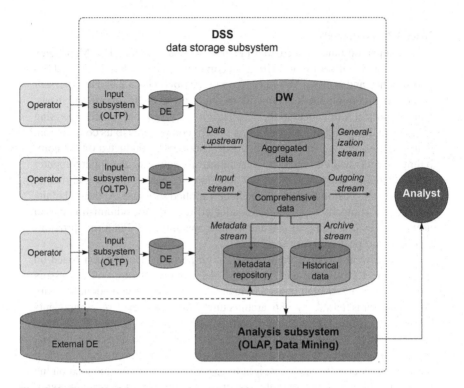

Fig. 3.22 Example of data storage subsystem architecture

The research groups from JSFC "Sistema" (OJSC "RTI") and Innopolis University are carrying out research and corresponding developments on cognitive dual-purpose supercomputers creation (Fig. 3.22). At this point, practice of engineering a cognitive system for the early detection of cyber-attacks showed that neural networks are sufficiently effective for attacks but at the same time are insufficiently informative as concerns explaining the possible methods of achieving these decisions. Whereas fuzzy logic systems were good at explaining their decisions, they were not suitable for automatically augmenting the system of rules required for decision-making.

To overcome these limitations, the decision was made to combine the abovementioned approaches to create an intellectual hybrid system, capable of handling cognitive uncertainties, as information security experts do. The created hybrid system belongs to the class of fuzzy neural networks. In this case, neural networks are used to configure membership in the fuzzy system that implements functions for the decision-making system for early cyber-attack detection. The fuzzy logic methods make it possible to directly describe the required scientific knowledge, using linguistic marking rules and neural-network learning techniques facilitated the automation of the design process and setting of the corresponding membership function that define these marks. Here the main stages of the computing process of early cyber-attacks detection are as follows:

- Development of fuzzy neural models based on biological neurons
- Development of synoptic connection models for introducing uncertainty into neural networks
- Development of learning algorithms (regulation method of synoptic weighting coefficients)

Modern data mining methods and models are based on statistical theory, fuzzy set theory, genetic algorithm, neural and cortical networks, etc. Neural networks proved to be quite efficient after passing the learning stage on the available data at detecting primary and secondary signs of cyber-attack. Here the network architecture (number of "layers" and number of "neurons" in each of them) should make it possible to solve early cyber-attack detection set problems with an acceptable computational complexity. In practice, choosing a similar network architecture was relatively difficult and involved a prolonged "trial and error" process. The designed neural network went through a lengthy training process, in which network neurons repeatedly processed the input data and adjusted their weights for better cyber-attack detection and prediction of further situation development. In this case, the practical results of cyber-attack detection, as well as accuracy of the corresponding predictions, were important. If only theoretically, neural networks can approximate any ongoing function of information confrontation. However, in practice, the final decision depends on the neural network's initial settings, and it cannot be interpreted in terms of known mathematical theories of phenomena and processes.

To detect "hidden" information on the early stages of preparing and planning cyber-attacks, it is possible to use data mining methods and models on structured and unstructured data. In this way, it is possible to perform:

- *Classification and regression* – determination of qualitative and quantitative value of the dependable object variable according to its independent variables
- *Search for associative rules* to detect frequent dependencies (or associations) between objects and events and *clusterization* – search for independent groups (clusters) and their characteristics among the dataset analyzed

The mentioned tasks are divided into descriptive and predictive (by designation) and into supervised learning and unsupervised learning (by solution methods).

In the future, it is planned to continue studies on the creation and development of a cognitive supercomputer with high performance for early cyber-attacks detection for critical infrastructure of the Russian Federation.

At the first stage (2017–2018), it is planned to create prototype models of a hardware and software artificial brain cortex of a living organism and develop a hardware and software package (PAC "Warning – 2016" [Rus. Preduprezhdenie-2016]) for early cyber-attack detection, including source environment development for independent association and synthesis of new knowledge concerning the quantitative and qualitative regularities of information confrontation (e.g., to synthesize scenarios of early prevention and opponent control in Russian cyberspace on extremely large structured and unstructured data volumes). Such artificial brain cortex prototype models should be capable of processing associative base types and solving the cognitive tasks of cyber-attack detection, warning, and consequence neutralization – from answers to new knowledge synthesis. Implementation of artificial brain cortex prototype models is supposed to be completed by developing modules of a corresponding cortical column (similar to the well-known "TASO Neuroconstructor" and IDE AI (Parallel IDE AI Neuro Standard, Neuro Sensorics, Neuro Genomics, Neuro CLoud, etc.) developments) which would make it possible to process ordered, partially ordered, and disordered associative bases, embedded and broken associative bases, etc.

At the second stage (2019–2022), it is panned to create a number of universal cognitive supercomputers with high (up to 10 PFLOPS) and ultrahigh (up to 1–3 EFLOPS) performance for early cyber-attack detection on critically important objects of Russian informational infrastructure. At the same time, it is expected to develop fundamentally new biomorphic neuron models with multiphase memory trace consolidation, cortical column models, cortical (similar to brain cortex) neural networks, and cybergenomic technologies for managing the growth and development of super-large artificial neural networks [4–10, 27–47].

At the end of the twentieth century, a crisis arose in mathematics, which manifested itself in increasingly complicated "pure" mathematics and its inevitable separation from applied mathematics (Fig. 3.23). The thing is that "pure"

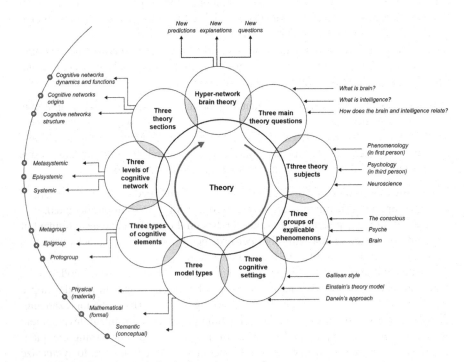

Fig. 3.23 Mathematics evolution in the early twenty-first century

mathematics operates with a high-level model of abstraction and a low-degree complexity (mathematicians often refer to this low-level complexity as *elegance*). In contrast, applied mathematics works with more specific models and simultaneously with higher levels of complexity (many variables, equations, etc.). That said, the majority of "interesting" applications of ideas of modern "pure" mathematics lies in the areas of *high abstraction and high complexity*. However, this knowledge area was previously almost inaccessible – mainly due to the human brain's limited ability to work with such models. Only after training computers to work with abstract mathematical objects did it become possible to expect that this problem will gradually lose its significance and we will begin to see interesting applications of abstract ideas in modern modern mathematics[23, 48–53].

In 2005, the talented Russian mathematician Vladimir Voevodsky, recipient of the Fields Medal award and professor at the Institute for Advanced Study in Princeton, proposed an approach to solve one of the main problems in the modern foundations of mathematics based on *homotopy theory* (parts of modern topology) and *type theory* (parts of modern programming languages theory). Informally, the abovementioned problem could be formulated in the following question form: "How can we correctly formalize the intuitive understanding that "identical" mathematical objects have "identical" properties? The well-known classical foundations of mathematics (in particular, the Zermelo-Fraenkel set theory) proved to be insufficient for the required formalization.

An international team of prominent researchers under the guidance of V. Voevodsky has developed a new foundation of mathematics called *Univalent Foundations of Mathematics* (Homotopy 2013 – Homotopy Type Theory: Univalent Foundations of Mathematics, Institute for Advanced Study, Princeton)[24] [53]. The practical value of the obtained results was that they made it possible to solve the problem of computer verification of mathematical proofs. According to Voevodsky, "these arguments push the set theory, as well as classical logic to the background. By doing so, the problem of an abstract mathematical language that computers will 'understand', can be considered as generally solved."

At the present time *Univalent Foundations of Mathematics* is an ongoing research project of the Institute for Advanced Study at Princeton. The objective is to build new axiomatic foundations of mathematics based on *homotopy type theory*. According to A. V. Rodin [54], "these new foundations, unlike the foundations constructed with standard logical methods, can be directly implemented in program code and as a result are closer to the goal set by Leibnitz" in his visionary Universal Characterization Project.

In this case, the homotopy type theory is an interpretation of Martin-Löf type theory by using geometric homotopy theory, which is an important part of modern algebraic topology. The difference between the "Univalent Foundations of Mathematics" and well-known axiomatic theories are the following. In the well-known Hilbert, Zermelo-Fraenkel, and other theories, logic is first defined (specifying its

---

[24]http://homotopytypetheory.org/book/

syntax and semantics), after which a formal logical system is used to formalize and axiomatize various informal theories and further study of different models of these theories. The same symbols, symbolic expressions, and syntactic operations with these expressions immediately have both geometric and logical interpretations in the homotopy type theory, which is a special kind of logical-geometric theory. For example, a *univalence axiom* is introduced geometrically as well as logically (generalization of the propositional Church extensionality principle). As a result, a straight conceptual connection exists between the Voevodsky approach and the development of Leibniz's idea of characteristics by Grassmann and Peano: in these cases, reference is made to formal symbolic calculus with semantics including both logical and geometric aspects. As A.V. Rodin notes [54], "the fact that homotopy type theory is not only a geometric but also a logical calculus, makes it possible to typify the primitive terms of this theory as basic material for constructing various mathematical theories. Taking into account how mathematical methods are used today in physics and other sciences, one expects that the given theory's geometric form will make it possible to apply this calculus more successfully in the sciences than the standard formal-logical methods."

It is essential that unlike most of logical methods, *homotopy type theory* permits immediate computer implementation (in particular, COQ and AGDA programming environment form), thus introducing the possibility of using this theory in modern information technologies for *knowledge representation* in some subject area. As a result, a programming language was developed to practically implement these ideas, which enabled the use of computers to verify mathematical arguments. Mathematicians once believed that abstract mathematics could not be reasonably formalized so accurately as to be "understood" by a computer. At the first research stage, which ended in autumn 2009, a computer was introduced to the ideas of *category* and *theoretic-homotopy* intuition, which serve as a basis for many modern mathematical constructions. Nowadays, more technical works are being completed to improve the language itself, in particular to solve the *Milnor* problem.

It is essential to note that the first examples of these languages were created in the late 1970s and were known as Martin-Löf type theories. The corresponding programming environments were created (e.g., COQ environment and programming language in France). However, there was no full understanding of the extent to which and how these languages can be used in practice. This is explained by the fact that only a small part of the language's possibilities were previously used based on the classical set theory. Against this backdrop, V. Voevodsky ideas fundamentally develop the functional capabilities of mathematical proofs and corresponding programming languages based on "homotomic" high complexity types. This is expected to significantly contribute to solving problems of how to apply "pure," modern mathematical ideas, including the solution of the mentioned practical cybersecurity problems.

References

1. Abramov, S.M.: Research in the field of supercomputer technologies of the IPS RAS: a retrospective and perspective. In: Proceedings of the International Conference "Software Systems: Theory and Applications", vol. 1, pp. 153–192. Publishing house "University of Pereslavl", Pereslavl (2009)
2. Guzik, V.F., Kalyaev, I.A., Levin, I.I.: Reconfigurable computing systems; [under the Society. ed. I.A. Kalyayeva], 472 p. Publishing house SFU, Rostov-on-Don (2016)
3. Kalyaev, I.A., Levin, I.I., Semernikov, E.A., Shmoilov, V.I.: Reconfigurable Multicopy Computing Structures; [under the Society. ed. I. A. Kaliayev]. 2nd edn, 344 p. Pub. House of the Southern Scientific Center RAS, Rostov-on-Don (2009)
4. Petrenko, A.S., Petrenko, S.A.: Super-productive monitoring centers for security threats. Part 1. Protect. Inf. Inside. **2**(74), 29–36 (2017)
5. Petrenko, S.A., Asadullin, A.Y., Petrenko, A.S.: Evolution of the von Neumann architecture. Protect. Inf. Inside. **2**(74), 18–28 (2017)
6. Petrenko, A.S., Petrenko, S.A.: Designing of corporate segment SOPKA. Protect. Inf. Inside. **6** (72), 48–50 (2016)
7. Petrenko, A.S., Petrenko, S.A.: Large data technologies (BigData) in the field of information security. Inf. Protect. Inside. **4**(70), 82–88 (2016)
8. Petrenko, A.S., Petrenko, S.A.: Super-productive monitoring centers for security threats. Part 2. Protect. Inf. Inside. **3**(75), 48–57 (2017)
9. Petrenko, S.A., Kurbatov, V.A., Bugaev, I.A., Petrenko, A.S.: Cognitive system of early warning about computer attack. Protect. Inf. Inside. **3**(69), 74–82 (2016)
10. Petrenko, S.A., Petrenko, A.S.: Lecture 12. Perspective tasks of information security. Intelligent information radiophysical systems. Introductory lectures [A. O. Armyakov and others; ed. S.F. Boev, D.D. Stupin, A.A. Kochkarova], pp. 155–166. MSTU them. N.E. Bauman, Moscow (2016)
11. Abramov, S.M., Lilitko, E.P.: State and prospects of ultra-high performance computing systems development. Inf. Technol. Comput. Syst. **2**, 6–22 (2013)
12. Khomonenko, A.D., Tyrva, A.V., Bubnov, V.P.: Complex of programs for calculation of reliability and planning of software tests. Federal Service for Intellectual Property, Patents and Trademarks: Svid. about the state. reg. software for the computer № 2010615617. Moscow (2010)
13. Khoroshevsky, V.G.: Architecture of Computing Systems. MSTU Them, 520 p. N.E. Bauman, Moscow (2008)
14. Kurnosov, M.G.: Models and algorithms for embedding parallel programs in distributed computing systems: Doctoral thesis in Technical. Science, 177 p. Siberian State University of Telecommunications and Informatics, Novosibirsk (2008)
15. Levin, I.I., Dordopulo, A.I., Kalyaev, I.A., Doronchenko, Y.I., Razkladkin, M.K.: Modern and promising high-performance computing systems with reconfigurable architecture. Proceedings of the international scientific conference "Parallel Computing Technologies (PaVT'2015)", Ekaterinburg, March 31–April 2, 2015, pp. 188–199. Publishing Center of SUSU, Chelyabinsk (2015)
16. Action plan. Document WSIS-03/GENEVA/DOC/5-R dated December 12, 2013. Geneva [Electronic resource]. Access mode: http://www.itu.int/dms_pub/itus/md/03/wsis/doc/S03-WSIS-DOC-0005*PDF-R.pdf
17. Active Engagement, Modern Defence. Strategic Concept for the Defence and Security of the Members of the North Atlantic Treaty Organisation adopted by Heads of State and Government in Lisbon. November 19, 2010 [Electronic resource]. Access mode: http://www.nato.int/cps/en/SID-14EF0623-198FC77E/natolive/official_texts_68580.htm
18. Advances in the field of information and telecommunications in the context of international security. Report of the UN Secretary-General. Document A/66/152 of 15 July 2011 [Electronic resource]. Access mode: http://www.un.org/en/documents/ods.asp?m=A/66/152

19. Abramov, S.M.: History of development and implementation of a series of Russian supercomputers with cluster architecture. In: History of Domestic Electronic Computers. 2nd edn, Rev. and additional; color. Ill.: Publishing house "Capital Encyclopedia", Moscow (2016)
20. Levin, V.K., et al.: Communication network MVS-express. Inf. Technol. Comput. Syst. **1C**, 10–24 (2014)
21. Wiener, N.: Cybernetics, or Control and Communication in Animal and Machine. 2nd edn, 344 p. Science, Moscow; The main edition of publications for foreign countries (1983)
22. Ashby, U.R.: Principles of Self-Organization, pp. 314–343. Mir, Moscow (1966)
23. Bongard, M.M.: The Problem of Recognition. Fizmatgiz, Moscow (1967)
24. Gavrilova, T.A., Khoroshevsky, V.F.: Bases of Knowledge of Intellectual Systems: A Textbook for High Schools, 384 p. Peter, St. Petersburg (2000)
25. Redko, V.G.: Evolution, Neural Networks, Intellect. LIBROKOM Book House/URSS, Moscow (2013)
26. Widrow, B., Stirns, S.: Adaptive Signal Processing. Radio and communication, Moscow (1989)
27. Petrenko, A.S., Bugaev, I.A., Petrenko, S.A.: Master data management system SOPKA. Inf. Protect. Inside. **5**(71), 37–43 (2016)
28. Petrenko, S.A.: Methods of detecting intrusions and anomalies of the functioning of cyber system, Proceedings of ISA RAS. Risk Manag. Safety. **41**, 194–202 (2009)
29. Petrenko, S.A.: Methods of ensuring the stability of the functioning of cyber systems under conditions of destructive effects. Proceedings of the ISA RAS. Risk Manag. Security, **52**, 106–151 (2010)
30. Petrenko, A.A., Petrenko, S.A.: Cyber units: methodical recommendations of ENISA. Quest. Cybersecurity. **3**(11), 2–14 (2015)
31. Petrenko, A.A., Petrenko, S.A.: Intranet Security Audit (Information Technologies for Engineers), 416 p. DMK Press, Moscow (2002)
32. Petrenko, A.A., Petrenko, S.A.: Research and Development Agency DARPA in the field of cybersecurity. Quest. Cybersecurity. **4**(12), 2–22 (2015)
33. Petrenko, A.A., Petrenko, S.A.: The way to increase the stability of LTE-network in the conditions of destructive cyber-attacks. Quest. Cybersecurity. **2**(10), 36–42 (2015)
34. Petrenko, A.S., Petrenko, S.A.: The first interstate cyber-training of the CIS countries: "Cyber-Antiterror2016". Inf. Protect. Inside. **5**(71), 57–63 (2016)
35. Petrenko, S.A.: Methods of Information and Technical Impact on Cyber Systems and Possible Countermeasures. Proceedings of ISA RAS. Risk Manag. Security, **41**, 104–146 (2009)
36. Petrenko, S.A., Petrenko, A.A.: Ontology of cyber-security of self-healing SmartGrid. Protect. Inf. Inside. **2**(68), 12–24 (2016)
37. Petrenko, S.A., Petrenko, A.S.: Creation of a cognitive supercomputer for the computer attacks prevention. Protect Inf. Inside. **3**(75), 14–22 (2017)
38. Petrenko, S.A., Petrenko, A.S.: From detection to prevention: trends and prospects of development of situational centers in the Russian Federation. Intellect Technol. **1**(12), 68–71 (2017)
39. Petrenko, S.A., Petrenko, A.S.: New doctrine as an impulse for the development of domestic information security technologies. Intellect Technol. **2**(13), 70–75 (2017)
40. Petrenko, S.A., Petrenko, A.S.: New doctrine of information security of the Russian Federation. Inf. Protect. Inside. **1**(73), 33–39 (2017)
41. Petrenko, S.A., Petrenko, A.S.: Practice of application of GOST R IEC 61508. Inf. Protect. Insider. **2**(68), 42–49 (2016)
42. Petrenko, S.A., Shamsutdinov, T.I., Petrenko, A.S.: Scientific and technical problems of development of situational centers in the Russian Federation. Inf. Protect. Inside. **6**(72), 37–43 (2016)
43. Petrenko, S.A., Simonov, S.V.: Management of Information Risks. Economically Justified Safety (Information technology for engineers), 384 p. DMK-Press, Moscow (2004)
44. Petrenko, S.A.: The concept of maintaining the efficiency of cyber system in the context of information and technical impacts. Proceedings of the ISA RAS. Risk Manag. Safety. **41**, 175–193 (2009)

45. Petrenko, S.A.: The Cyber Threat model on innovation analytics DARPA. Trudy SPII RAN. **39**, 26–41 (2015)
46. Petrenko, S.A.: The problem of the stability of the functioning of cyber systems under the conditions of destructive effects. Proceedings of the ISA RAS. Risk Manag. Security. **52**, 68–105 (2010)
47. Petrenko, S.A., Kurbatov, V.A.: Information Security Policies (Information Technologies for Engineers), 400 p. DMK Press, Moscow (2005)
48. Aristotle. Comp. in 4 volumes (Series "Philosophical heritage"). Thought, Moscow. (1975–1983)
49. Biryukov, D.N.: Cognitive-functional memory specification for simulation of purposeful behavior of cyber systems. Proc. SPIIRAS. **3**(40), 55–76 (2015)
50. Bocharov, V.A., Markin, V.I.: Fundamentals of Logic. Moscow State University, Moscow (2008)
51. Chereshkin, D.S.: Problems of Information Security Management, 224 p. Editorial URSS, Moscow (2002)
52. Grinyaev, S.N.: The battlefield – cyberspace: theory, methods, means, methods and systems of information warfare, 448 p. Harvest, Jordan (2004)
53. Voevodsky, V.: Voevodsky V. A Very Short Note on the Homotopy Lambda-Calculus (2006)
54. Rodin, A.V.: Logical and geometric atomism from Leibniz to Voevodsky. Prob. Philos. **6** (2016)

References

...

Chapter 4
Possible Scientific-Technical Solutions to the Problem of Giving Early Warning

4.1 Possible Problem Statement

This chapter investigates the complex issue of an early-warning system for cyber-attacks on Russian state and corporate information resources. An approach to create the required warning systems based on "computing cognitivism" is proposed; it is a relatively new scientific research area with cognition and cognitive processes being a kind of symbolic computation. It is shown that the cognitive approach makes it possible to create systems that are fundamentally different from the traditional systems for cyber-attack detection, prevention, and recovery (*SOPCA*). *SOPCA* has a unique ability to independently associate and synthesize new knowledge on qualitative characteristics and quantitative patterns of information confrontation. A feasible architecture of a cognitive early-warning system for a cyber-attack against Russian information resources based on convergent nano-, bio-, info-, and cognitive technologies, *NBIC* technologies is proposed [1–5].

Starting in 2005, technologically developed countries witnessed an evolutionary transition from classic supercomputer models in the spirit of von Neumann to a new model of the "computer of the future," based on the organizational structure of a living organism's brain (i.e., models of artificial cognitive computational systems). The fundamental difference between this new supercomputer model and that of its predecessors lies in the properties of self-organization and adaptivity demonstrated in the system's capacity for prolonged learning and behaving independently in real-time computing conditions.

This section considers the fundamental questions of designing an artificial cognitive system to solve tasks related to giving early warning of cyber-attacks. Analysis is first provided of the cognitive approach's historical results (from Aristotle to Kolmogorov and Marr) before giving consideration to the fundamental principles and particularities of designing supercomputers intended to give early warning of cyber-attacks [6–12].

© Springer International Publishing AG, part of Springer Nature 2018
S. Petrenko, *Big Data Technologies for Monitoring of Computer Security: A Case Study of the Russian Federation*, https://doi.org/10.1007/978-3-319-79036-7_4

4.1.1 Historical Background

Recently, several developed countries (USA, EU countries, China, Russia, etc.) have witnessed a new technological structure of society on the basis of "convergent" NBIC technologies. For example, in the USA, the National Science Foundation and the US Department of Commerce program run the *NBIC* program (nanotechnology, biotechnology, information technology, and cognitive science). Similar *GRAIN* programs (genetics, robotics, artificial intelligence, and nanotechnology) and *BANG* (*bits, atoms, neurons, genes*) are being implemented in the European Union. In China, a similar program "China Brain" was launched. In Russia, the national technological initiative *NeuroNet* (*CoBrain* program or Web 4.0) was launched [13–17] to produce experimentally hybrid and artificial bio-like materials, technical bionic-type systems, and technological platforms based on them. This initiative was implemented by some of the leading Russian research institutes and universities, including *RTI*, Russian research center *Kurchatov Institute*, the *Kogan Research Institute of Neurocybernetics*, *Mozhaysky Military Space Academy*, MIPT, St. *Petersburg Electrotechnical University*, *National Research University of Infor mation Technologies*, *Mechanics and Optics*, etc. There are future plans to develop complex anthropomorphic technical systems and "nature-like" technologies, designed to combine animate and inanimate components of nature.

The first results of NBIC technology research, as well as the known "computational cognitivism" results, have given rise to a fundamentally new type of cybersecurity system – the "artificial cognitive systems" [8, 18, 19] – which could be used to create a cognitive early-warning system for cyber-attacks on corporate and state information resources of the Russian Federation. This system is able to cognize and behave independently in the context of information confrontation. For instance, it allows solving the following relevant issues:

- Image identification (patterns and clusters), recognizing the preparation, and implementation of cyber invasion
- Standard scenarios for prevention, detection, and counteraction in cyberspace development
- Generation, accumulation, and processing of new knowledge on the quantitative patterns of information confrontation
- Presentation of the "deep" semantics of information confrontation
- Preparation and implementation of response decisions which are relevant to the specific cyber-attack, etc.

The term "cognitive" comes from the Latin word "cognitio" meaning cognition. The cognitive approach is first mentioned in Ulric Neisser's monograph *Cognitive Psychology* [20], where the brain activity processes are considered as information processing ways. The further development of mathematical models to represent mental processes boosted the development of the cognitive approach in the technical field. The first "artificial cognitive systems" represented "intelligent" software and hardware systems based on the traditional von Neumann architecture [8, 18, 19, 21, 22].

The modern cognitive approach is based on methods of information cognition, perception, and accumulation, as well as methods of thinking or using this information for "judicious" problem solution. The artificial cognitive systems are considered to be able to "repeat" the complex behavioral functions of the nervous system and even human thought processes. This, in fact, explains the interest in the cognitive approach for solving the problem of an early-warning system for cyber-attacks on Russia's critically important information resources. Let us consider a possible problem statement (Inset 4.1)

Inset 4.1 Possible Problem Statement
The component part of the experimental work (CP EW) is called "Development of a cognitive system for an early-warning system for cyber-attacks" (CP EW software-hardware complex [SHC] "Warning 2016").

Purpose: SHC "Warning 2016" is intended for an early-warning system for cyber-attacks on corporate and state information resources of the Russian Federation.

Sample content: SHC "Warning 2016" should include the following subsystems.

Data collection and knowledge subsystem designed to collect structured and unstructured information of qualitative characteristics and quantitative patterns of information confrontation based on big data technologies. At the same time, corporate Internet/Intranet systems; connected telecom operators data transmission networks; industrial Internet of things (IIoT) systems and Internet of things (IoT) in general; social and media systems, i.e., Facebook, Instagram, Twitter, LinkedIn, VKontakte, Odnoklassniki; and also the following messengers, Skype, Viber, WhatsApp, Telegram, FireChat, etc., could be considered as information resources.

Big data preliminary processing and analysis subsystem designed to identify the facts of planning, preparation, and implementation of cyber-attacks in real (quasi-real) time in the early stages of computer aggression based on various artificial intelligence methods (including identification of useful knowledge based on signature, correlation and invariant methods of pattern recognition, network traffic analysis algorithms, cognitive finite-state machines and agents, self-learning neural networks, etc.).

Data and knowledge storage subsystem designed to provide long-term storage of incoming and processing information, including cognitive methods. This subsystem should provide detection of various security threat types and primary and secondary signs of computer aggression in the background.

Modeling, preparation, and decision-making subsystem designed to prepare and make response decisions preventing the forced transfer of protected infrastructure to catastrophic states. The following new NBIC models are to be supported:

(continued)

Inset 4.1 (continued)
– Neuromorphic, similar to living nervous system structure
– Corticomorphous, similar to the cerebral cortex structure
– Genomorphic, similar to genetic and epigenetic reproduction mechanisms and living organisms development

Also the following models and methods of mathematical logic and artificial intelligence should be supported:

- Cognitive agents
- Artificial neural networks of direct distribution, trained by the Levenberg-Marquardt method
- Trainable, hierarchically ordered neural networks and binary neural networks
- Various representations of dynamic thresholds and network packets classifiers based on Euclidean-Mahalanobis metrics and the support vectors method
- Statistical (correlation) and invariant profilers
- Complex poly-model representations, etc.

Subsystem of visualization and administration designed to display the results of the created complex, the preparation, and the making of appropriate decisions, as well as administration and monitoring of the created complex (including on the basis of developed geoanalytics and geoinformation systems).

4.1.2   Cognitive Approach Prerequisites

A significant contribution to the formation and development of the modern cognitive approach was made by many leading domestic and foreign scientists working on various scientific fields:

- Theoretical informatics and artificial intelligence
- Special computer science and artificial intelligence issues (i.e., ontology, pseudophysical logic, spatial reasoning models)
- Dialog systems models and methods
- Fuzzy sets, linguistic variables, and information granulation theory
- Agents and multi-agent systems
- Intelligent robots and robot control systems
- Special issues of cybersecurity and the functioning stabilization of the national infrastructure critical information systems ensuring

The cognitive approach is based on the following fundamental results:

- Mathematical logic (from Aristotle to A. Kolmogorov)
- Mathematical theory of computability (from A. Turing to A. Maltsev)
- The computing machines developed by von Neumann architecture
- The generative grammars theory by N. Chomsky
- The computational neurophysics theory by J. Marr, etc.

"Computational cognitivism" was formed by the following outstanding scientists:

Aristotle introduced the main terms and definitions of logic, the concept of logical operation, and defined the basic laws of thought, including the laws of contradiction and third exclusion [23]. In so doing, he attempted to rationalize by connecting the act of calculation to the initial assumptions. As a result, he came very close to one of the sections of future mathematical logic named the theory of evidence.

G. *Leibniz* proposed a way to translate the logic "from the verbal kingdom, full of uncertainties, into the realm of mathematics[169]." In other words, he suggested that scientists move from fruitless disputes to calculations in order to prove their rightness.

J. Buhl developed a system of notations and rules applicable to all kinds of objects, from numbers and letters to sentences, which made it possible to formulate assertions as statements, the truth or falsity of which could be proved [2]. As a result, he laid the foundations of modern mathematical logic.

A. *Turing* introduced the scientific algorithm concept. Simultaneously, equivalent schemes were proposed by A. Church and E. Post [2].

The further development of classical propositional logic (*J. Boule, G. Frege*, and *B. Russell*) and the mathematical theory of computability (*A. Turing, A. Church*, and *E. Post*) made it possible to formalize the ideas of *T. Hobbes and G. Leibniz* that cognitive processes could be represented as computational and algorithmic processes.

A. *Newell* and *G. Simon* introduced the concept of a physical symbolic system representing the realization of a universal machine "producing a changing set of symbolic structures with time [46]."

A. *Chomsky* proposed the generative grammar theory [2, 24, 25], in which fundamental differences were made between the language competence (the language's knowledge of the ideal-speaking listener) and the real use of language, by the deep and surface levels of syntactic representation, and postulated the presence of congenital grammatical structures.

D. *Marr* proposed a universal scheme for analyzing the information processing with natural and artificial technical systems.

A. N. *Kolmogorov* fundamentally developed such concepts as algorithm, automaton, randomness, entropy, information, and complexity. Moreover, he put forward the most important methodological thesis of artificial life: "if the property of a particular material system" to be alive "or to have the ability to" think "is defined in a purely functional way (for example, any system with which one can discuss

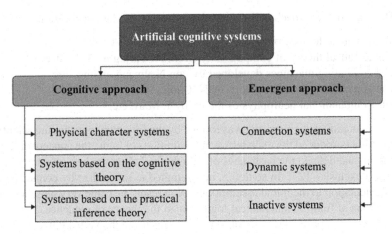

Fig. 4.1 Feasible classification of cognitive artificial systems

problems of modern science and literature will be recognized as thinking), then it will be necessary to recognize, in principle, the feasible creation of living and thinking creatures" [26].

Contemporary studies of cognitive systems are conducted based on the neuro-physiological principles present in the structure of the nervous system construction and the methods of human cognition and mental activity. For instance, in [27–30], the application of artificial cognitive systems with hybrid architectures in robotics is justified. In this case, a cognitive system is defined as a system able to learn and adapt to its environment and to change it due to the accumulated knowledge and skills acquired in the functioning process. Two main types of artificial cognitive systems are clearly distinguished: cognitive and emergent (Fig. 4.1).

Cognitive systems include:

- Traditional character systems (*Newell and Simon*)
- Systems based on the knowledge theory using symbolic knowledge learning and acquisition (*Anderson*)
- Systems based on the practical inference theory and high-level psychological concepts of persuasion, desire, and intention (*Bratman*)

Here, the first are able to generate some symbolic structures or expressions. In this case, a symbol is a physical pattern representing some expression (or character structure) component. The second ones are based on a system of production and a generalized thinking model and person cognition containing memory, knowledge, decision-making, training, etc. At the same time, the training contains declarative and procedural stages, depending on the knowledge of the trainee. The rest implement the decision-making process, similar to the traditional practical conclusion.

Emergent systems:

- Connection systems
- Dynamic systems
- Inactive systems

The first ones implement parallel processing of distributed activation patterns using statistical properties, rather than logical rules. The second ones study various self-organizing motor systems and human perception systems, analyzing the corresponding metastable behavior patterns. In the third, the definition of the cognitive essence is purposeful system behavior, which occurs when it interacts with the environment.

Thus, a general methodology for developing hybrid cognitive engineering systems is proposed [11, 30].

1. Formalized cognitive concepts and methods for creating effective self-learning and self-modifying systems.
2. Methods for the synthesis of original cognitive components (modules and module networks) capable of accumulating knowledge through learning and self-learning. In this case, the components are based on a combination of neurological, immunological, and triangular adaptive elements – the most effective for multidimensional functional approximation, as well as the corresponding behavioral networks.
3. Methods for cognitive components and systems implementation based on specially developed software. The software implementation of cognitive components is based on the processing of original information and training models and cognitive systems based on multi-agent technology. Here, the cognitive multi-agency makes it possible to develop the distributed cognitive systems with a high level of behavior complexity.

In [27, 31–34], an ontologies system for cognitive agents is presented using mobile robots as an example. An information granulation meta-ontology is constructed. The ontologies representation is given based on fuzzy algebraic systems. Spatial relations have been explored for the interactive management organization of cognitive mobile robots. A meteorological approach to the analysis of spatial ontologies is described. Clear and fuzzy mereotopological spatial relations are considered. A granular space ontology is proposed. It is noteworthy that the results [33, 34] are also suitable for solving the problem of early detection of cyber-attacks on the Russian information resources.

4.1.3 Technological Reserve for Problem Solution

As a technological basis for solving this problem, it is proposed to consider modern software and hardware systems for analyzing and processing information security events. In international practice, these complexes are developed as part of specialized security centers, known as the *Computer Emergency Response Team* (*CERT*) or the *Computer Security Incident Response Team* (*CSIRT*) or the *security operations center* (*SOC*).

The Russian Federation has already established a number of state and corporate centers for detecting, preventing, and recovering from cyber-attacks or centers for responding to cybersecurity incidents, which are similar to foreign *CERT/CSIRT/ SOC* in their functionality. In domestic practice, they are known as *SOPCA*. Some examples include, inter alia, *GOV-CERT.RU* (*FSS* of Russia), *SOPCA of the Ministry of Defense of Russia*, *FinCERT* (Bank of Russia), *Rostechnologies CERT*, *Gazprom SOC*, etc.

The Russian Federation Presidential Decree No 31c of January 15, 2013 "On the establishment of a state system for detecting, preventing, and recovering from cyber-attacks on Russian information resources" establishes that the Russian FSS is making methodological recommendations on the organization of protection of the critical information infrastructure of the Russian Federation and organizes work on the creation of a state and corporate segments of monitoring in the detection, prevention, and cybersecurity incident response (SOPCA).

The conceptual framework of a state system for detecting, preventing, and recovering from cyber-attacks on Russian information resources No K 1274, approved by the President of the Russian Federation on December 12, 2014, lays down a system overview of a state SOPCA system, which is based on special centers for detecting, preventing, and recovering from cyber-attacks. These are in turn subdivided into the following:

- Russian *FSS* (created to protect information resources of the public authorities)
- State and commercial organizations (created to protect their own information resources)

In addition, these centers are coordinated by the *National Coordinating Center for Computer Crimes* under the *FSS* of Russia.

At the same time, in practice, the task to develop a cognitive early-warning system for cyber-attacks on the information resources of the Russian Federation was far from being trivial. It was necessary to conduct appropriate scientific research and solve a series of complex scientific and technical problems – e.g., input data classification, identifying primary and secondary signs of cyber-attack, early cyber-attack detection, multifactor prediction of cyber-attacks, modeling of cyber-attack spread, training, and new knowledge generation on quantitative patterns of information confrontation – many of which did not have ready standard solutions. In addition, it was essential to ensure the collection, processing, storage of *big data*, as well as carrying out analytical calculations on extremely large amounts of structured and unstructured information from a variety of *Internet/Intranet* and *IoT/IIoT* sources (*big data* and *big data* analytics) [8–12, 14–16, 18, 19, 21, 22, 33–48].

A possible list of requirements for such cognitive systems is represented in Inset 4.2

Inset 4.2 SOPKA System Requirements
While implementation of SHC "Warning 2016," there were a number of general requirements:

- Monitoring a large number of objects number real time (1000000+)
- Low delay level in event processing (less than 10 ms)
- Distributed storage and fast access to data for petabyte data volumes
- A high-reliability degree of data and knowledge storages able to operate 24 hours a day, 7 days a week, without risk of interruption or loss of information in the event of server failure (one or more)
- Ability to scale (including the means of the underlying software) for the performance and volume of processed information without modifying the installed software by upgrading/scaling the used set of hardware
- Indicator of the SHC availability level should be at least 99% per year
- Possibility of SHC integration with third-party systems: the complex architecture should be created, taking into account the openness and ease of introducing interaction modules with external systems

Creating the data storage of SHC "Warning 2016" is implemented taking into account the following requirements:

- Data, stored in the repository, is a series of records characterized by a time stamp (time series); thus, the repository should be optimized for storing time series.
- High speed of data recording.
- High speed of MapReduce operation with preliminary selection on time intervals.
- Ability to work with data with a coordinate as one of its properties (mobile sensors).
- Low requirements for data consistency (eventually consistency).
- Immutable data, without the need to conduct distributed transactions or synchronization.

Requirements for data and knowledge collection, preliminary processing, and analysis subsystem:

- Receiving data on various information interaction protocols, i. e., ZMQ (zeromq), TCP/IP, RAW TCP/IP, HTTP (REST-requests processing), AMQP, SMTP, etc.
- Receiving data in XML, JSON/BSON, PlainText formats
- Detection of incidents and security threats by applying the following models to the incoming data stream:

(continued)

Inset 4.2 (continued)
- Various parameters excess/decrease detection, setting thresholds for these parameters
- Detection of deviation from normal values for various parameters
- Detection of statistical deviations from standard behavior for various parameters in the time window and average value change over a certain time interval
- Change in characteristics of events occurrence frequency, etc.

- The use of machine learning models to identify correlations and detect incidents (i.e., the application of multivariate analysis, clustering and classification methods)
- Identification of various kinds of templates in text messages described by regular expressions and applying the abovementioned statistical functions to them
- Correlation of data from various sources
- Combination of parameters from various sources with subsequent application of the abovementioned statistical methods
- Testing of the models for detecting new incidents, etc.
- Giving notification to users of incidents detected by sending messages to the visualization and administration subsystem or other IS

Requirements of the data storage and knowledge subsystem:

- Support for structured and unstructured data types
- Support for data index to speed up data search and retrieval
- Ability to work with time series
- Ability to create queries in the MapReduce paradigm
- Implementation of aggregation and statistical queries on the time series in the data storage location
- Ability to automatically remove outdated data of time series
- Availability of libraries for accessing storage functions for Java and Python
- Library for accessing storage system functions through specialized drivers (e.g., Django database engine)
- Knowledge support for working with new models of neurophysics and classical methods of artificial intelligence

The modeling, decision-making, visualization, and administration subsystem needed to support models and methods of neurophysics, artificial intelligence, and mathematical logic, including cognitive agents and artificial neural networks of direct distribution, trained by the Levenberg-Marquardt method, and so on.

The visualization and administration subsystem needed to support:

(continued)

Inset 4.2 (continued)
– Statistical reports on incidents and stored time series
– Density distribution function graphs
– Cybersecurity values distribution histograms
– Series graphs with different characteristics (mean, extrapolation, etc.)
– Correlation models for performing multifactor analysis
– Parameters correlation and incidents on selected time interval graphs
– Classification models to detect correlation of parameters from various sources with incident occurrence
– Clustering models for detecting parameters correlation over a given time interval, etc.

Appropriate technological solutions for creating a cognitive early warning for cyber-attacks on Russia's information resources are represented in Fig. 4.2.

Here, the choice and implementation of the *big data* processing component represented an important task (Table 4.1).

Another important task was the structure of big data storage structure. Many known solutions (e.g., *Cassandra* or *HBase*) proved to be of little use due to the following limitations:

- Lack of database components to ensure efficient storage and retrieval by time series (most known solutions do not contain integration tools due to their closeness and those available (e.g., *InfluxDB*) do not have a high level of work stability)
- Absence of the logical connections between the interfaces of business logic and the database
- System functionality duplication due to the database and the processing logic being separated in a heterogeneous solution environment
- Limited performance of the *HBase* solution, associated with the architectural solution features
- Significant overhead *Cassandra*, associated with the synchronization of data on various nodes, etc.

Possible system architecture of the cognitive early-warning system for cyber-attacks on information resources of the Russian Federation based on NBIC technologies is presented in Fig. 4.3.

The positive experience gained in the creation of a cognitive early-warning system for cyber-attacks of SHC "Warning 2016" speaks to the expediency of a methodical approach to solving the task.

Stage 1 Developing the technical component of a traditional *SOPCA* based on big data technologies is the creation of a high-performance corporate (state) segment of detecting, preventing, and recovering from cyber-attacks.

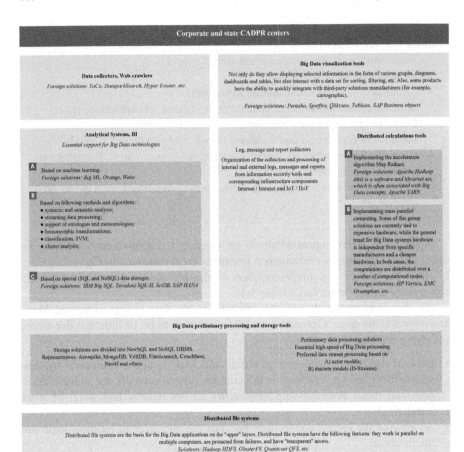

Fig. 4.2 Technological reserve for problem solution

Table 4.1 Known solutions for streaming and batch data processing

Solution	Developer	Type	Description
Storm	Twitter	Packaged	New solution for big data streaming analysis by Twitter
S4	Yahoo!	Packaged	Distributed streaming processing platform by Yahoo!
Hadoop	Apache	Packaged	First open-source paradigm MapReduce realization
Spark	UC Berkeley AMPLab	Packaged	New analytic platform supporting data sets in RAM has high failure safety level
Disco	Nokia	Packaged	MapReduce distributed environment
HPCC	LexisNexis	Packaged	HPC-cluster for big data

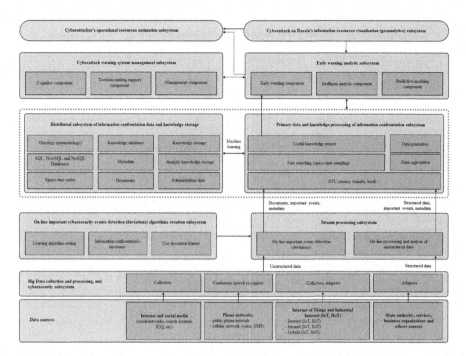

Fig. 4.3 Possible architecture of the cognitive early-warning system for cyber-attacks

Stage 2 Creation of the SOPCA analytical component based on "computational cognitivism" is the realization of the cognitive component of the cyber-attack early-warning system capable of independently extracting and generating useful knowledge from large volumes of structured and unstructured information for SOPCA operational support.

In this case, the above mentioned technical component of *SOPCA* based on big data technologies should be appropriately allocated with the following functions:

• Big data on the information security state in controlled information resources collection
• Data detection and recovery after cyber-attacks on information resources
• Software and technical tools for IS events monitoring support
• Interaction with the state *SOPCA* centers
• Information on the detection, prevention, and recovery from cyber-attacks, etc. (Fig. 4.4)

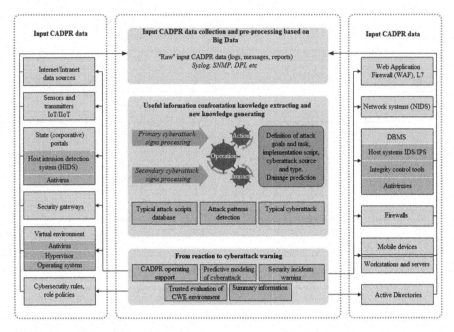

Fig. 4.4 Technical component of SHC "Warning 2016"

The analytical component based on "computational cognitivism" should be appropriately allocated with the following functions:

- An early-warning system for cyber-attacks on information resources
- Identification and generation of new useful knowledge about qualitative characteristics and quantitative patterns of information confrontation
- Prediction of security incidents caused by known and previously unknown cyber-attacks
- Preparation of scenarios for deterring a cyber-opposition and planning a response and adequate computer aggression (Fig. 4.5).

In the following sections on this issue, the practice of using big data technologies to organize a streaming process of cybersecurity data, as well as practical questions of semantic master data management (MDM), will be considered for building the SOPCA knowledge base. The development of a new functional model of a cognitive high-performance supercomputer will also be justified, possible prototypes of software and hardware complexes for the early detection and prevention of cyber-attacks will be presented, and examples of solutions to classification and regression problems will be given, as will be solutions to the search for associative rules and clustering and possible directions for the development of artificial cognitive cyber-security systems.

Cognitive subsystem of early warning system for cyber-attacks on Russian information resources

1. Development of traditional SOPCA components based on big data technologies

2. Creation of analytical SOPCA component based on "computational cognitivism"

3. Monitoring the security and sustainability state of critically important infrastructure operation in cyberspace:

- introducing the informatization CWE passport (inventory, categorization, classification, definition of requirements, etc.);
- creating priority action plans, etc.;
- certifying the information security tools (facility attestation for safety requirements);
- monitoring safety criteria and indicators and stability of the given facilities' operation;
- maintaining a database of cybersecurity incidents.

4. Identification of preliminary signs of cyber-attacks on Russian Federation information resources:

- recognizing structural, invariant, and correlation features of cyber-attacks;
- adding primary signs of cyber-attacks to the database;
- clarifying cyber-attack scenarios;
- developing adequate measures for deterrence and compensation.

5. Identification of secondary signs of cyber-attacks on Russian Federation information resources:

- identifying correlation links and dependencies between the signs;
- adding secondary signs of cyber-attacks to the database;
- clarifying cyber-attack scenarios;
- developing adequate measures for deterrence and compensation.

6. From detection to prevention:

- early warning for a cyber-attack on Russian information resources;
- prediction of cyber - attack from a cyber enemy;
- assessing possible damage in case of cyber-attack;
- preparing scenarios of deterrence and coercion response to the cyberworld preparation

7. Extraction of useful knowledge and generation of new knowledge in the field of information confrontation based on:

- new NBIC models:
- neuromorphic, similar to the living nervous system structure;
- corticomorphous, similar to the cerebral cortex structure;
- genomorphic, similar to genetic and epigenetic mechanisms of living organisms' reproduction and development;
- models and methods of mathematical logic and artificial intelligence;
- cognitive agents;
- artificial neural networks of direct distribution, trained according to the Levenberg-Marquardt method;
- educable, hierarchically ordered neural networks and binary neural networks;
- various representations of dynamic thresholds and classifiers of network packets based on the Euclidean-Mahalanobis metric and the support vectors method;
- statistical (correlation) and invariant profilers;
- complex poly-model representations, etc.

8. Development of guidelines for work with cognitive SOPCA

9. Cyber-training organization to develop skills of early warning for cyber-attacks on information resources of the Russian Federation

10. Development of the necessary normative documents

11. Training and retraining of employees on issues relating to the early warning for cyber-attacks on information resources of the Russian Federation

12. Elaboration of proposals for the development of a national (and international) regulatory framework for cyber-attack early warning.

Fig. 4.5 SHC component "Warning 2016"

4.2   Applying Big Data Technology

The applicability of big data technologies for domestic information security services is explained by the need to extract and use valuable knowledge from large structured and unstructured information in the future [8, 18, 19, 31]. Here, information sources can be corporate Internet/Intranet systems, industrial Internet and Internet of things (*IIoT/IoT*), social and media systems (*Facebook, Instagram, Twitter, LinkedIn, VKontakte, Odnoklassniki*), instant messengers (*Skype, Viber, WhatsApp, Telegram, FireChat*), etc. For this reason, the term *big data* is one of the most frequently used terms on various information platforms and cybersecurity forums.

4.2.1 Introduction

Nowadays, technologies for processing, storing, and analyzing *big data* are becoming increasingly applicable for monitoring the security state of the Russian critically important infrastructure (electrical networks, oil pipelines, communication systems, etc.) as well as monitoring relevant criteria and indicators of information security and operation sustainability in general.

Big data technologies are already being used in a number of cybersecurity applications. For example, non-relational databases (*NoSQL*) are used to store logs, messages, and security events [8, 19, 21, 31] in security information and event management (*SIEM*) systems. In the near future, a qualitative leap in *SIEM* development is expected based on models and methods of predictive analytics. In the known solutions of *Red Lambda, Palantir*, etc., *big data* technologies are used to build user profiles and social groups in order to detect abnormal behavior. At the same time, corporate mail, *CRM* system and staff system, *access monitoring and control system (AMCS)*, as well as various extractors in the affiliate network Internet/ Intranet and *IIoT/IoT* data transmission networks, external news feeds, collectors, and aggregators in social networks are used as information sources.

The applicability of such technologies (Fig. 4.6) is further confirmed by the possibility in principle to analyze online data (both packet and streaming), to select and process significant simple and complex cybersecurity events in real (or quasi-real) time scales, and to generate new useful knowledge for detecting and giving warning of security incidents. *Big data* can also provide proactive security and monitor impending information security incidents before they adversely affect the sustainability of critical infrastructure.

Thus, by *big data information security technologies*, we mean technologies of efficient processing of dynamically growing data volumes (structured and unstructured) in heterogeneous Internet/Intranet and *IIoT/IoT* systems for solving urgent

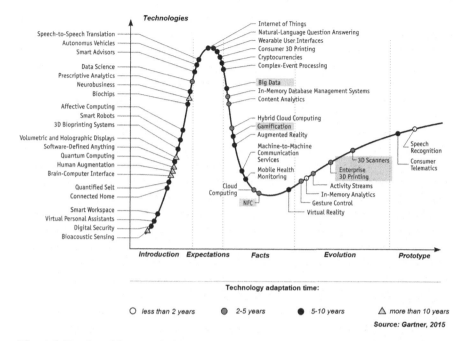

Fig. 4.6 Trends and future evolution for the information technology

security tasks. The practical significance of *big data* technologies lies in the ability to detect primary and secondary signs of preparation to conduct a cyber-attack, to identify abnormal behavior of monitored objects and subjects, to classify previously unknown mass and group cyber-attacks (including new *DDOS* and *APT*), and to detect traces of cybercrime, etc. (i.e., when using traditional information security tools [*SIEM*, *IDS/IPS*, information security tools from unauthorized access, data encryption tools, antiviruses, etc.] is not very effective).

It should be expected that *big data* technologies will significantly expand the functionality of classical systems and information security tools by seamlessly supplementing them with new functions for analyzing and processing *big data*. For example, it is virtually impossible to obtain reliable analytics from traditional SOC storage or the transactional *SIEM* system about data streaming every hundredth of a second. However, this task becomes solvable by using suitable *big data* technologies. In other words, *big data* technologies make it possible to solve the problems of preliminary analysis, storage, and processing of structured and unstructured information and reliable and unreliable (or unknown [*Dark Data*] – incomplete and fuzzy) data in real (quasi-real) time scale. They also provide an opportunity to extract and generate new useful knowledge in information security based on artificial intelligence and mathematical logic, parallel computations and informatics theory, image recognition and electronic detection (*eDiscovery*) methods, genetic and evolutionary algorithms, self-learning neural and multi-agent cognitive networks, etc. (Fig. 4.7).

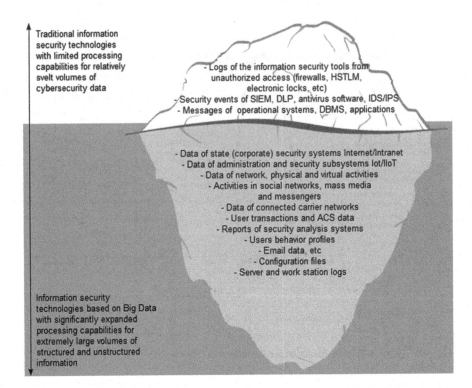

Fig. 4.7 Difference between traditional and new approaches to information security based on *big data* technologies

4.2.2   *Big Data Comparative Analysis*

At present, the known approaches for data stream processing [11, 18, 21, 31, 41] are based on:

- Classic *CEP* model (e.g., *StreamBase*; *MapReduce* modifications, e.g., *D-Streams*)
- Actor models (*Storm*, *S4* and *Zont*)
- Combinations of the actor model and MapReduce modification (*Zont + RTI*)

In the first approach, the classic *complex event processing* (*CEP*) model is applied (Fig. 4.8). *CEP* enables searching for "meaningful" cybersecurity events in a data stream over some time interval, analyzing event correlation, and detecting the appropriate event patterns that require immediate response.

For automation the process of developing the data stream processing systems based on *CEP*, a number of tools are proposed (e.g., the *StreamBase* development environment with its own declarative programming languages *StreamSQL* and *EventFlow*). A sample problem representation in the *EventFlow* language is shown in Fig. 4.9.

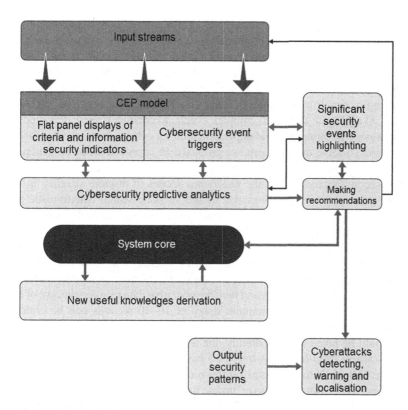

Fig. 4.8 Sample CEP architecture

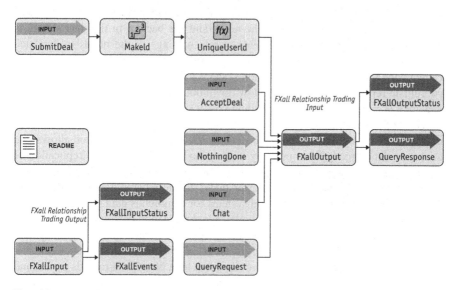

Fig. 4.9 Sample description in the EventFlow language

The languages mentioned above are equivalent in their expressiveness and contain typical data input/output procedures, text and graphic job processing, statement outputs in terms of the corresponding relational algebra, etc. In addition, these languages support the basic constructions of programming languages Java, R, etc.

The practice of using *CEP* has shown that it is optimal for the collection and processing of simple cybersecurity events – e.g., to extract events from several data streams, to aggregate them into complex events, to conduct reverse decomposition, etc. However, the complex logic implementation for cybersecurity events processing is difficult. To solve this problem, we proposed an approach based on generalizing MapReduce for processing the data stream.

The second approach uses the *D-Streams* model of discrete streams, in which streaming computations are presented as sets of nonessential deterministic packet computations on small time series intervals. It is significant that such a representation of computations made it possible not only to implement the complex logic of processing cybersecurity events but also to offer better methods of restoration than traditional replication and backup copying. In practice, computer networks with a large number of nodes (from hundreds or more) inevitably experience failures and "deadlocks" (or "slow" nodes), and here the operative data recovery in case of failure is quite important: even a minimum delay of 10–30 seconds can be critical for making a correct decision.

It should be noted that, apparently, the well-known systems for processing data streams (*Storm, MapReduce Online*, etc.) have reached the threshold values of fault tolerance. All of them are based on the model of "long-lived" session operators, which, upon receiving the message, update the internal state and send a new message further. In this case, the system is restored by replicating it to a preprepared copy of the node or by backing it up in a data stream, meaning it "replays" the messages on each new copy of the "fallen" node. Using the replication mechanism results in a costly double/triple reservation of the node, and the backup in the data stream is characterized by significant time delays while waiting for the nodes to "update" while the data is reprocessed through the operators. In addition, these approaches cannot cope with deadlock. Replication systems use *flu* synchronization protocols to coordinate replicas, and deadlocks slow down both replicas. When backing up, any deadlock is considered a failure entailing costly subsequent recovery.

The D-Stream model offers more advanced recovery methods (Fig. 4.10). For example, the *resilient distributed datasets (RDD)* data recovery method allows you to restore data directly from memory without losing sub-seconds on replication or the parallel state restoration method of a "lost node" in which a "drop" of a node initiates connecting the working cluster nodes to the recomputation of the "lost" *RDD* structure. In contrast, traditional systems for continuous data processing do not permit such recovery due to their complex synchronization protocols.

In practice, the D-Stream model is implemented in the spark streaming data processing system (Fig. 4.11), which is advantageous for its tactical and technical characteristics. In particular, it can process more than 60 million records per second at 100 nodes with sub-second delays. In case of failure, system recovery also occurs in sub-second time. In general, the spark streaming performance in terms of one node

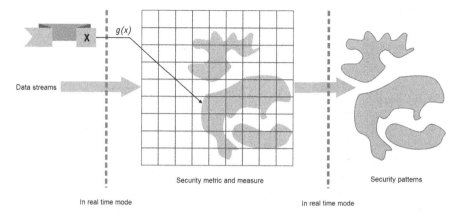

Fig. 4.10 Clustering example

Fig. 4.11 Spark streaming algorithm

is comparable to commercial streaming databases, while it can scale linearly up to 100 nodes and run 2–5 times faster than the open-source systems *Storm* and *S4*.

The *D-Streams* model requires splitting the array of input data into streams, which inevitably leads to partial event loss. In addition, in the case of large flows, the data processing system is no longer flexible and scalable. The system's response time to events slows down, and the system moves further away from the real-time mode.

To solve these problems, the **third** of the abovementioned approaches was proposed, based on the actor model. Here, "actors" refer to parallel computation entities. The main advantage of actors is the ability to store states, including those obtained from historical data which can be used to highlight significant cybersecurity events. Among the known solutions for streaming data processing, based on the actor model, are the *Storm (Twitter)*, *S4 (Yahoo!)*, and *Zont (MIPT)* systems.

The first two of these solutions [11, 18, 21, 31, 33, 41] permit the implementation of pipeline data processing based on a relatively small number of actors. The development of these solutions is done with Java programming language applying the *JMS* machine, which is used to transmit asynchronous messages between *finite-*

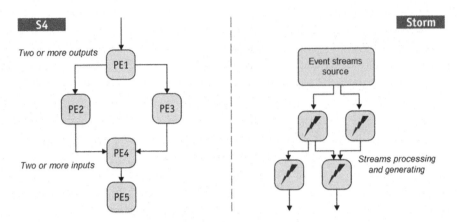

Fig. 4.12 Differences in the S4 and Storm topologies

state machines (*FSM*). In S4, the program is defined in terms of processing elements graph and in Storm in graph terms from Spouds and Bolts. Here, Spouds are sources of event streams, and Bolts implement the software logic for processing and generating flows (Fig. 4.12).

The third of the named system – *Zont* – makes it possible to work with a large number of actors. This is important when working with a sensor cloud, when one actor is assigned to each. To develop distributed fault-tolerant systems, it is possible to use the *Erlang* and *RIAK Core* development environment. Here, the functional language *Erlang* (*Ericsson*) can be used to develop programs that can work in a distributed computing environment on several nodes (processors, cores of one processor, cluster of machines), and the open library *Riak Core* (*Basho Technologies*) permits the development of distributed applications according to *Amazon's Dynamo* architecture.

Thus, all three systems (Storm, S4, and *Zont* [*MIPT*]) can be used to process a large data stream from several sources. In this case, *Zont* is optimally fitted for working with a sensor cloud.

The **fourth** approach combines the advantages of the second and third approaches, which allows creating real-time data stream processing systems based on a combination of the MapReduce modification and the actor model.

4.2.3 Sample Solution Based on Big Data

Let us consider possible options for building a cognitive early-warning system for cyber-attacks on Russian information resources (SHC "Warning 2016") based on *big data* technologies [11, 18, 21, 31, 41].

Version 1 SHC "Warning 2016" Exploratory Prototype Implementation

The basis of the proposed solution was the non-relational distributed database *HBase*, working on top of the *Hadoop Distributed File System (HDFS)*. This database can perform analytical and predictive operations on data terabytes to assess cybersecurity threats and the overall stability of critical infrastructure. It is also possible to prepare, in an automated mode appropriate detection, neutralization and warning scenarios.

The analytical component of the PAK "Warning 2016" exploratory prototype consists of two modules: hypothesis generation and hypothesis analysis. The system of generating hypotheses is distributed between the server and client module of hypotheses and events with event information being stored in *HBase*. This predictive module's results are hypotheses about the optimal algorithms for detection, neutralization, and prevention.

The second module of hypothesis analysis is designed for processing large volumes of information; therefore, high performance is required. The module interacts with standard configuration servers and is implemented in *C* language (via *PECL* [*PHP* extensions repository]). Special interactive tools based on *JavaScript/CSS/DHTML* and libraries such as *jQuery* have been developed to work with the content of the proper provision of cybersecurity.

The exploratory prototype includes standard and special configuration servers and databases based on *PHP*, *MySQL*, and *jQuery* with *SSL/HTTPS* protocols being used for communication. At the same time, *EnReal* interacts with servers using *REST* (*representational state transfer*) protocols via secure http channels.

MySQL Percona Server (version 5.6) with the *XtraDB* engine is used for data storage. The database servers are integrated into a multi-master cluster using the *Galera Cluster*. *Haproxy* is used for balancing the database servers. *Redis* is involved to implement task queues, as well as data caching (version 2.8).

A *nginx* is used as a web server. *PHP-FPM* with *APC* enabled is involved. *DNS* (multiple *A-records*) is used for balancing http requests.

The programming languages *Objective C*, *C++*, and *Apple iOS SDK* based on *Cocoa Touch*, *CoreData*, and *UIKit* are applied to develop special client applications running *Apple iOS*. The mentioned applications are compatible with devices running *Apple iOS* version 9 and higher. The native Google SDK is applied to develop applications running *Android OS*.

These applications are compatible with running devices *Android OS* version 4.1 and higher. Software development for the web-based platform is carried out using *PHP* and *JavaScript*.

The SHC "Warning 2016" exploratory prototype will be deployed to *LETI* on the *DigitalOcean* platform and contains:

- Three servers for the *production stage*
- One server for the *testing stage*

In addition, *Cloudflare* is used to increase the service speed (through the use of *CDN*) and protection against DoS attacks.

The load testing conducted on the SHC "Warning 2016" exploratory prototype indicates the viability of the proposed technical solution [7, 49–54].

Version 2 The implementation of SHC "Warning 2016" (MIPT) exploratory prototype and its possible architecture are shown in Fig. 4.13.

The telematics platform *Zont* served as the basis of the proposed solution, which makes it possible to create fault-tolerant scalable cloud systems for *big data* stream processing. Table 4.2 describes the modules of the SHC "Warning 2016" exploratory layout based on *Zont*.

It is significant that *Zont* has its own specialized storage, built on the basis of *Riak Core* technology, for storing and retrieving archived data which represent time series. For the repository backend, *LevelDB* is used, which is an embedded *KV* database (under development at Google) designed to be used as a *backend* in the construction of specialized databases and provide the operations of writing, searching, and sequential data viewing. *DB's* chief advantages include a high speed of data recording, predictable speed of data search by key, and high speed of sequential reading. It also worth noting the following important characteristics of *LevelDB* for the storage of time series:

- The use of the LSM tree model, making it highly resistant to failures
- Ordered data storage

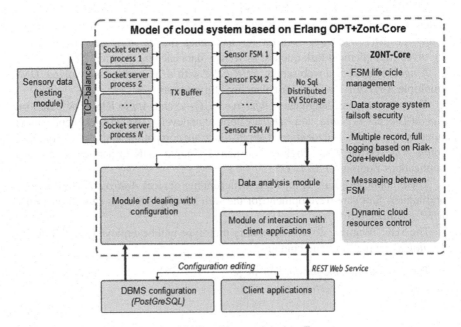

Fig. 4.13 Possible SHC "Warning 2016" architecture based on Zont

Table 4.2 Description of data processing modules

Element (name in the schema)	Element functionality
TCP balancer	Distribution of the network load between cluster nodes
TCP session manager (socket server process)	Different network connection management for each sensor
	Preliminary consistency check of input data
Transactional buffer (TX buffer)	Buffering input data to optimize peak performance
	Data parsing procedure call from sensors
	Data redirection to the corresponding FSM
Sensor finite-state machine (sensor FSM)	Selected FSM process for each sensor
	Logical data processing from sensor
	Event tracking within a single detector
	Saving processed data to a database
Distributed KV storage (NoSql distributed KV storage)	Saving processed data
	Improved reliability of recording and reading data due to repeated data storage on different nodes
Data analysis module	Data analysis, complex events tracking
	Generation of reports based on database content
Module of interaction with client applications	Interaction with external clients using REST and WebService

Table 4.3 Technical specification of stand equipment

Tech specs/cluster	Standard network cluster	High-performance cluster I-SCALARE
Number of servers for software "platforms" mock-up	6	200
Processor	2 × 2 Intel Xeon E5–26206-core 2,0 GHz	2 × 2 Intel Xeon E5–2690 (8-core) 2,9 GHz
Cache	32 Gb	64 Gb
Hard disk drives	3 Tb	600Gb (SATA)
Netware	Ethernet 1Gb	InfiniBand – QDR (40 Gb/s)

In addition, to support the input data of mobile sensors, geo-index support based on geohash technology was added. This index is specifically designed to store a large array of information about the spatial position of point features, in particular, moving sensors, and yields better performance indicators than other well-known universal solutions as Postgres *GiST* and *MongoDB 2dsphere*.

The hardware implementation of the SHC "Warning 2016" exploratory layout is a general-purpose server cluster, united by a network.

The technical specification of the stand equipment is presented in Table 4.3 and a typical scheme of hardware implementation in Fig. 4.14. The testing device includes the following components:

- Virtual machine server *Erlang* (*Erlang node server*) to implement the distributed cloud platform

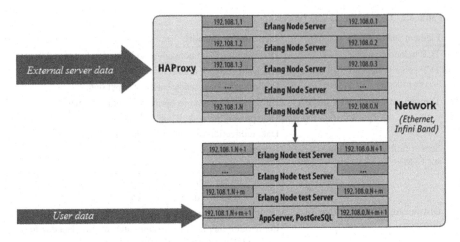

Fig. 4.14 Software and hardware implementation of the stand

- Virtual machine server *Erlang* (*Erlang node test server*) for test module implementation, including the sensor network emulator
- Server of the auxiliary software "platforms," containing the application server JBoss and the DBMS *PostgreSQL*

The system and application software of the testing device included:

- OS Linux CentOS 6.4
- Erlang R160B2 as the execution environment of the distributed machine Erlang
- HAProxy – as a proxy server
- JBoss – as an application server
- PostgreSQL – for storing metadata, etc.

Preliminary tests of the experimental design of the PAC "Warning 2016" demonstrated its potential [11, 18, 21, 31, 33].

The positive experience obtained in applying *big data* to solving information security problems attests to the expediency of choosing solutions based on the actor model and MapReduce technology over distributed *KV* cloud storage. At the same time, solutions based on *CEP* turned out to be less demanding in terms of memory, since they permit storing data in a single "window" of events. Nonetheless, they required significant computing resources for analyzing such "windows." Solutions, based on the model of actors, proved less demanding for computational resources but more demanding for memory because of the need to duplicate the data for each event/object. Accordingly, the solutions based on the modification of *MapReduce* took an intermediate position.

In our opinion, *big data* technologies can dramatically change the state of affairs in the following areas of information security:

- Early cyber-attack detection, prevention, and recovery
- Proactive management of cybersecurity incidents

- Prognostic network monitoring
- User authentication, authorization, and identity management
- Prevention of computer crime and fraud
- Information security risk control
- Control regulatory compliance, etc.

In this case, the first results should be expected precisely in the proactive management of cybersecurity incidents and early warning of cyber-attacks, which has also been confirmed by the results of foreign developers of information protection tools. For example, in 2015–2016, *RSA* and *IBM* announced strategies for creating a new generation of *security operations center (SOC)* called *intelligence-driven security operations center, iSOC*.

4.3 Feasible Models and Methods for Giving Warning

At the present stage of the information security tools development, a real need has arisen to create systems to prevent and give early warning of cyber-attacks on critically important information resources of the Russian Federation. These systems should take a worthy place within the state system of detecting, preventing, and recovering from cyber-attacks (*SOPCA*) [1, 2, 11, 18, 21, 31].

4.3.1 Introduction

In [1], a conceptual model was proposed for proactive behavior during an information and technical conflict (*ITC*) in a heterogeneous network infrastructure using anticipation. The idea of assigning cyber systems to the ability of preemptive behavior during *ITCs* seems very promising, since in most cases it is preferable to prevent impact of attacks on *critically important information infrastructure (CIII)*, rather than deflect them and even more to neutralize their consequences.

The proactive cyber system behavior in the conflict is proposed [55] to be reduced to the synthesis of such a behavior scenario, during which it is able to thwart the adversary's planned attack, thus averting negative consequences for *CIII*. At the end, the cyber system must be able to change the course of one of the events that is an integral process part which would be carried out to the detriment of the *CIII* (cyber system). A performance change in one of the activities should lead to a change in the process, namely, a transition to a sequence of events (to a process trajectory/track) with a satisfactory outcome for the *CIII* (cyber system).

Therefore, it was proposed to frame the research problem relating to issues of ensuring *CIII* security during *ITC* as one of building a system capable of generating scenarios of anticipatory behavior in a conflict [56], in which the anticipation property [55–59] should be realized, opening to it the ability to act and make concrete decisions with a certain temporal-dimensional anticipation regarding the expected future events.

4.3.2   The General Appearance of Anti-cyber Systems to Prevent the Cyber Threat Risks

In order to provide information security cyber system with property of anticipation, the following capacities are presumed:

- Receiving information through the sensor system
- Handling information about past system experience
- Comparing information received with that available
- Hypothesizing about possible future events
- Generating strategies for targeted system behavior
- Maintaining the required *CIII* security level

The work [59] proves that the desired system can be a self-learning intellectual system [1, 60] of self-organizing [2] gyromates [1, 55, 58, 61]. Also, it introduces additional system requirements for generating scenarios for preemptive behavior during conflict: (1) when implementing the "reasoner" in the projected system, it is necessary to consider the possibility of applying it to "open worlds," and, consequently, (2) the fact base representing the conflict domain should be capable of representing an "open world."

The advanced requirements (1, 2) owe to the fact that the designed system should be potentially capable of forming anticipatory behavior scenarios in new conflict types (scenarios requiring prevention which were not directly included into the system during its creation).

Requirement (3): The "synthesizer" must choose behavior strategies that are adequate not only to the information environment state but also to the system goals, which may change during its operation.

If we consider different goals as different contexts which influence the decisions made, then we can assume that the context should influence the solutions availability.

In general, the desired cyber system should be able to detect potentially dangerous processes (i.e., "tasks") and find suitable "solutions."

In the initial design stages for cyber systems for preventing cyber-attacks with the capacity to anticipate, analysis was conducted on the most studied features of human memory and its working functions [59]. This is because precisely humans are able to synthesize scenarios of preemptive behavior at different levels by using various mechanisms from the nervous system in general and the brain in particular.

The main element of the system for early possible attack detection and its preventive suppression is the module for scenario synthesis of anticipatory behavior in the information and technical conflict — the gyromate. The system itself is a partially ordered hierarchy of gyromates with level-by-level coordination [59], and it makes it possible to solve the consistency problem in the model completeness conditions of the theory underlying the designed system. Each particular gyromate must consist of four basic elements: the interpreter, the planner, the generator, and the memory. The memory plays a critical role because the global and local interaction of the first three (basic) elements depends upon it (Fig. 4.15). Requirements for the memory of the designed system are defined in [59].

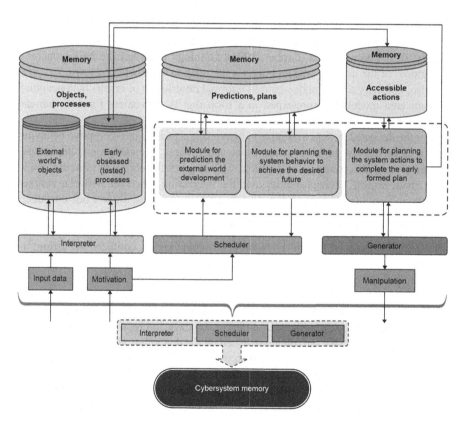

Fig. 4.15 Organizational design of cyber system memory

4.3.3 Proposals for Knowledge Representation for an Intelligent Risk Prevention System

Considering the data management features in human semantic memory [56, 58, 59, 62], it is advisable to represent knowledge about the domain area (*DA*) in the form of a hierarchy of structured objects connected by relationships. This idea is based on such representation models as frames, semantic networks, *UML*, etc. Unfortunately, all these languages, despite being convenient means for presenting knowledge on the *DA* conflict, cannot reflect semantics, and all information that they express is intended for human [rather than machine] perception. Also, first-order logic (*FOL*) has developed greatly; there are mechanisms of semantic processing, but there are no convenient means for representing knowledge.

In 1935, R. Brahman proposed that combine semantic networks and *DA* be combined. The result was called terminological logic, and then, in the development process, it turned into the descriptive logic family [1], which found its practical application in describing ontologies in the currently developing semantic web.

Ontology is a formal conceptualization specification that takes place in a certain context of the domain area (Gruber, 1993). Ontology describes the basic concepts (positions) of the domain area and defines the relationships between them.

Also, ontology can be understood as a formal description of results in conceptual domain area modeling, represented in a form that is convenient for both human perception and implementation in a computer system [49].

For the designed system to be able to synthesize scenarios of preemptive behavior in conflicts, it must be able to represent and process knowledge on the causes, processes, and consequences of conflicts. This requires the ability to present knowledge about objects, their properties, and the interaction processes between various objects and entities. However, the domain areas in question should not introduce restrictions that lead to the impossibility of describing knowledge from these *DA* in a single ontology. Otherwise, it can lead to the principle impossibility of enriching the *intellectual system* (*IS*) with knowledge of behavior in conflicts occurring in other *DAs*.

An important issue with knowledge manipulation is represented through a single ontology in the *IS knowledge base* (*KB*), associated with the allocation of those fragments from ontology, the knowledge of which should be available to the system when solving particular problems in different contexts. In Fig. 4.16, each concept is a certain action type (A); the role represented by the dotted line should be interpreted as "a subclass," and the role depicted by a solid line "follows behind." If we draw an analogy with the mechanisms that take place in human memory, then this problem can be reformulated as the task of controlling the focus of attention [55] and the task of placing part of the data into the episodic buffer [58] for further processing.

Considering this, multiple requirements were formulated [59], which should be taken into account in creating a meta-modeling system for representing and manipulating knowledge. The knowledge metamodeling system should have the following capabilities:

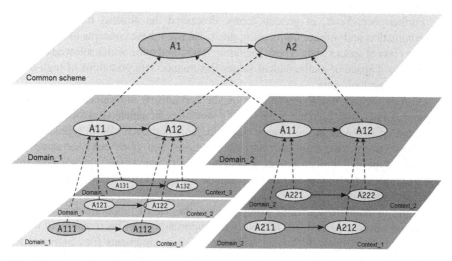

Fig. 4.16 A sample common context-dependent on domain area of knowledge representation in a single ontology

- Represent and process knowledge from different domain areas about their objects, properties, and processes (for the possibility of enriching the *KB* with descriptions of various conflicts that may be observed in various *DA*).
- Contain a limited number of concepts and role types, sufficient for describing an arbitrary *DA* (for setting unified rules for building an ontology that further consolidates them, as well as applying unified logical rules for generating new knowledge regardless of the *DA* specifics and the conflicts).
- Have solvable algorithms for directional data retrieval (to realize the possibility of determining the necessary ontology fragment based on tasks and contexts).
- Search by analogy for similarity of the ontology fragment (to perform a search by analogy of similarity with arising tasks and their solutions during conflicts in different *DAs*).
- Change the knowledge availability for further processing (to order the obtained solutions, to identify the most associated knowledge, and to "forget" false and underutilized/weakly confirmed knowledge).

The desired abilities require a formal semantic definition of the denotational knowledge contexts in the ontological modeling of the conflict domain areas.

In order to enable the manipulation of abstractions, it is necessary to have a suitable, mathematically defined language. In [63–65], D. Scott proposed a noteworthy view on the nature of data types and functions (mappings) from one data type to another. He also pointed out that the questions connected with the description of functions, in mathematics, defined in all admissible functions as arguments and applicable even to themselves as to the argument, were not sufficiently worked out and proposed a mathematical theory of functions that can be used as a project that gives a "correct" approach to semantics.

Furthermore, Scott, in general terms, described the abstract theory of finite approximation and infinite limits using grids. He used these concepts to effectively build a class of spaces (data types), including functional ones, which allowed them to be used as a space of mathematical values in semantic interpretations of high-level programming languages.

A. Shamir and W. Wage, relying on the formalisms introduced by D. Scott, presented a new approach to the data type's semantics, in which types themselves are included as elements in the objects' domain. This approach allows types to have subtypes , to examine truly polymorphic functions, and gives exact semantics for recursive type definitions (including definitions with parameters). In addition, this approach provides simple and direct methods for proving the typical properties of recursive definitions.

In the modeling of semantic computations, the use of functional types is very promising. Scott regarded [63–65] as a very important part of his research, in which he examined functions, for which functions serve as arguments (i.e., higher-order functions). He was one of the first to use functional space as a semantic structure, but many of his colleagues did not understand him at that time. Indeed, however, it is the functional spaces that should be taken into account when selecting and implementing operators (based on the simulated processes semantics) that perform mappings on functional spaces. Such operators are nothing more than combinators (in terms of H. Curry and A. Church) or functional forms (in terms of G. Backus).

To simplify the perception and use of functional types in practice, it is useful to be able to explicitly specify their partially ordered set in the form of appropriate type concepts, as well as the relationships between them (Fig. 4.17). This should facilitate the ontological modeling process of the conflict domain areas.

Based on the analysis results, we can assume that any data type used in ontological modeling of conflict domain areas can be described through the list of (1) the objects that correspond to it, (2) the properties of these objects, and (3) the action functions (and for common case) that can be performed by the objects being considered [49].

If we consider abstract classes of objects O, properties P, and actions A, then as a whole, they can be represented in the grid form containing one lower element, one upper element, and three discretely separated elements located on the same "latitude."

If we take into account the fact that an object can consist of lower-hierarchy objects (simpler objects) and be part of higher-hierarchy objects (structurally more complex

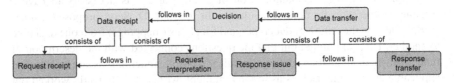

Fig. 4.17 Sample of explicit assignment of functional types and relations between them

objects), then in general it is possible to talk about structures, not objects. However, for the sake of simplicity in presentation, it is further proposed to use the term "objects," thereby denoting a certain "section of observation" under certain contexts.

In order for the designed cyber system to become truly intelligent, it must be able to structurally store data, knowledge, and its contextual binding to tasks and solutions and also perform complex cognitive actions ("think logically") to generate new knowledge that is not directly included in the cyber system but acquired while setting and solving possible tasks. The proposed approach to ontological data representation in the intelligent system memory, based on type theory and finite approximation theory, opens the feasibility of a formal presentation of context-dependent, multilayered knowledge about domain areas, potentially containing knowledge of the conflicting parties' behavior strategies. In the future, this approach should make it possible to discover the correct procedure for performing plausible reasoning over context-dependent knowledge presented in the memory of the system for preventing cyber threats [8, 18, 19, 33, 59].

In [59], it is proposed to give the intellectual system (*IS*) the possibility of conceptual and logical transformations over data and knowledge in the evolution of its multifunctional memory. In this regard, it is required to find the possibilities of logical extension, constructive typification, and semantic aggregation of the incoming data flow, facts, and knowledge with already accumulated knowledge represented in the ontological layers of semantic memory [59]. The latter are stored as the set of associated "semantic links" (*SL* are two concepts associated with the role) [11, 41, 59]; its main types are represented in Table 4.4, where *Fn* $[X \rightarrow Y]$: the designation of the relations between the concepts of type X and the concepts of type Y. The semantics of using relations are represented in the field "*SL* description" of the table below.

Table 4.4 Main types of semantic links for arbitrary domain area ontologies

SL denotation	SL description
F11: $[P \rightarrow P]$	«Property» approximates «Property»
F12: $[O \rightarrow O]$	«Object» approximates «Object»
F13: $[A \rightarrow A]$	«Action» approximates «Action»
F2: $[O \rightarrow P]$	«Object» has «Property»
F3: $[P \rightarrow A]$	«Property» produces the ability to «Action»
F4: $[A \rightarrow P]$	«Action» is suitable for the impact at «Property»
F5: $[A \rightarrow A]$	«Action» follows in «Action»
F6: $[O \rightarrow O]$	«Object» consists of «Object»
F7: $[A \rightarrow A]$	«Action» consists of «Action»
F8: $[O \rightarrow O]$	«Object» produces «Object»

The whole *SL* set is a generalized ontology of those domain areas where IS operates. At the same time, absolutely all concepts included in the various *SL* and presented in the ontology must be hierarchically linked and belong to one of the partially ordered sets in the basis *{objects, O; properties, P; actions, A}* [1, 55–62, 66].

When a new fact is received at *IS* input, its knowledge system links the received data with the previously constructed ontology by a special means of formal-logical inferences.

4.3.4   General Approaches to Knowledge Generation by an Intelligent System

To simulate plausible conclusions about the data represented in the ontology, an approach based on an extended syllogistics application is proposed.

In syllogistics, various logical relationships between categorical attributive statements are explored. In the composition of categorical attributive statements, quantifier words that predicate bundles and terms are distinguished. In each categorical attributive statement, there are two terms: *subject (a)*, a term denoting those objects about which a statement is either confirmed or denied, and *predicate (b)*, a term denoting what is being predicated, confirmed, or denied about these subjects.

Categorical attributive statements, depending on the number, are divided into single and multiple. Among the multiple, there are common ($\forall$ and private ($\exists$) statements. Depending on the quality, the discussed statements are divided into affirmative and negative.

Implementing the mechanism for performing plausible inferences, it is necessary to take into account the restrictions imposed by the semantics of the roles used (see above or [59]).

Considering the semantic constructions based on *SL* and containing the role "consists of ...," it should be taken into account that in this case, the division procedure for subjects [59] (objects, processes) is described, and not the restrictions of concepts scope, which leads to the impossibility of using direct syllogistic modes.

For further description of the order of arguments simulated, based on inference with semantics roles, the notion of primitive semantic construction (*PSC*) suggests the inference feasibility. *PSC* is a pair of the semantic links that have one common concept in their composition.

It can be argued that the consideration of the rules for constructing arbitrarily complex ontologies, as well as the rules for the implementation of plausible inferences on knowledge represented in the form of an ontology (through the use of semantic links, the types of which are listed above and in [59]), can be reduced to the consideration of rules for constructing primitive semantic constructions, as well as consideration of the rules for the implementation of plausible reasoning over them (since it is the *SL* pair that can participate in generating new knowledge). If we take into account that (1) each *SL* is characterized by two concepts and the role between them and also that (2) in case of the interaction between two CL, they should have one common concept (the case where two concepts coincide in pairs is not considered, since it does not imply the feasibility of making a logical conclusion), you can build a table with possible conceptual triplets. Since only three types of concepts are proposed for constructing an ontology ("objects," "properties," and "actions" [1, 55–62, 66]), then a table of possible combinations of them, consisting of triple concepts, will also be small – only 27 variants. In this case, each option requires a certain consideration, since it is necessary to consider all feasible (admissible) roles and their directions, as well as the combinations of quantifiers.

One of the notable and important features of the developed knowledge system is its ability to generate concepts and incorporate them into an already existing ontology.

So, if the IS needs to compute the value of a function at a point outside the definition area, then a new syntactic object is generated at that point, which reflects the call fact at a given point. This object in [8–12, 14–16, 18, 19, 21, 22, 33–48, 59] is suggested to be defined as a string that completely coincides with the call text. This way of defining functions is called a delay in the function computation or simply a delay in a function. The value resulting from such a delay is suggested to be called an intensional function (for concision, an intensional).

The analysis of widely used inference machines (FaCT ++, HermiT, Pellet, etc.) and the descriptive logics implemented in them [42, 59] showed that they are unable to automatically form new concepts, to name and to integrate them into a pre-existing ontology.

The positive effect of computations over the extended area is particularly marked in the modeling of multistage processes (e.g., multistage attacking impacts and the ways to protect against them). At the same time, it should be noted that during the computation delay, a new data type, represented as an intensional, is generated.

Initially, an arbitrary domain area concepts are represented in the ontology in the simple object terms form, while the concepts names obtained in the course of the intensional expansion belong to the class of applicative object terms. An essential property of applicative object term names is that they record their own "history" of appearance. The noted terms can be used further in constructing the simulated processes specifications.

Thus, the proposed mechanism for processing facts coming into the intelligent system input is aimed at their automatic integration into the existing ontology, thus generating new knowledge. New knowledge generation is carried out in the course of constructing conclusions, based on a simple categorical syllogism, taking into account the semantics of the roles used and also in the process of replenishing the ontology with the intensionals generated. Both of these methods are implemented in the knowledge systems (an example of such being the gyromate) and are performed in interrelation with each other, which makes it possible in the course of precomputing not only to supplement the ontology with completed roles but also to integrate new, generated concepts into the ordered set of concepts. Ontology replenishment is automatically performed by the gyromate and significantly increases its information content and significantly reduces the time required for replenishment.

However, generating new knowledge by applying these methods is not enough to synthesize process specifications that describe strategies for proactive behavior in a conflict. During the synthesis of the specified specifications, it is proposed to additionally use the developed method of establishing the semantic similarity of the "tasks" and "solutions" models by analogy. A. Poincaré, R. Courant, G. Robbins, D. Poya, and I. Lakatos noted in their works that analogy plays an important role in mathematical creation.

The method kernel for establishing the semantic similarity of the "tasks" and "solutions" models by analogy is based on the feasibility of the gyromate in the general case to make the transition from approximating (specific) concepts (and consequently, processes) belonging to one *DA* by way of approximable (more general, generic) to approximating but belonging to another *DA* [16, 19, 34, 42, 43].

Admittedly, it is not always possible to find the "tasks" and their "solutions" using new knowledge derived only by analogy, since the intelligent system may not contain the knowledge necessary. For this reason, it seems useful to realize in the "synthesizer" the possibility of generating new knowledge by combining existing ones.

Arbitrary combining can lead to a "combinatorial explosion" and to the generation of absurd process models; therefore, a mechanism for directional combining to form models of potentially realizable processes should be implemented in the *IS*. For this purpose, it is suggested to use the capabilities of the apparatus of applicative-combinatorial computations [8, 19] that facilitate the realization in the gyromate of the procedure of directional combining, based on the appliance of a limited list of available functional forms $[[O_1 \rightarrow P_1] \rightarrow A_1 \rightarrow [O_2 \rightarrow P_2]]$, which means "an object O_1 possessing property P_1 is able to effect the impact of A_1 on an object O_2, since the latter possesses property P_2." This makes it possible to reduce the number of incorrect specifications generated at the stage of their formation. In other words, the gyromate can take the decision on the potential implementability (nonrealizability) of the process, described by the generated specification on the basis of the knowledge extracted from the constructed ontology, since it contains data on the types of objects (in the form of the concept grid) and functional types (in the form of a roles grid). This information allows the gyromate to correctly combine the available combinators (as functions) and data (as arguments).

In general cases, the complex application result of these methods are gyro-generated specifications, suitable for organizing processes to prevent possible SC. Obviously, the specification expansions can be different and include specific programs for managing specific agents, sensors, and effectors.

The observed events, information about which is entered into the knowledge base, do not always have a strictly probabilistic nature (real functions on the space of these events do not satisfy Kolmogorov's three axioms). In addition, it is necessary to take into account that experts do not always operate with clear information and use in their reasoning not only quantitative but also qualitative categories.

This attests to the increasing number of difficulties that arise in solving the problems of forming specifications for the processes of preventing new types of destructive impacts on critically important information infrastructure.

Taking the abovementioned information into account, we propose to pay attention to approaches that allow us to develop a kind of "superstructure over ontology" or a system of accounting preferences that are, in effect, subjective in nature [8, 19, 59].

4.3.5   Feasible Models and Methods for Preempting

The [1, 2] suggests a network model that allows structuring the memory in a way that information retrieval becomes possible by following the associative chains that are automatically created and changed at the network topology level – in the process of receiving and processing information (requests). The used information is often more accessible, and the power of associative links increases with the number of entities mentioned together. Such storage and retrieval of information in memory partially imitate Hebb's cellular ensembles [59]. As it seems, this approach to knowledge ordering in memory can be applied when implementing procedures for storing and processing data in the memory of an intelligent cybernetic system.

In [1, 2, 42], a memory model, called an associative resource network, is described. This model is represented by an oriented graph with a variable topology. The vertices correspond to the domain area entities and the edges to the associative links between them. The proposed model fits well with the model of ontology representation of domain areas.

Each entity represented in the associative resource network has brightness. Brighter vertices indicate more visibility or more availability for searchers. Edges have a limited capacity. The more often two vertices participate in a query together, the bigger the capacity of edges that connect these vertices. The bigger the edge capacity is between two vertices, the bigger the associative force between the relevant entities (vertices). The associative resource network itself (ARN) is built and operates over the domain area ontology (Fig. 4.18) [19, 42, 44], which is being built and used in the intelligent activity of the cyber system, designed to solve the applied tasks of information security.

$$v[id] = v_{\langle id, tv, brt \rangle} \text{ is the ARN vertex,}$$

where $id = NV(v_{\langle id, tv, brt \rangle})$ is the unique vertex identifier (name, which can correspond to the name of the ontology concept with which the ARN vertex is linked and can be represented, for example, as a number $id \in ID\_V$, where ID_V – set of ARN vertices identifiers)

$tv = TV(v[id])$ is the type of ontology concept with the considered ARN uniquely linked $tv \in \{O, P, A\}$, where O is <objects>, P is <properties>, and A is <actions>)

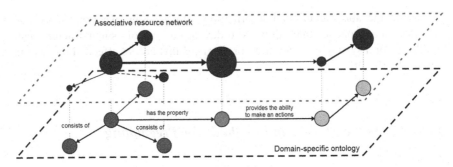

Fig. 4.18 Diagram of the joint functioning of the associative resource network and the domain area ontology

$brt = BV(v[id])$ is the "brightness" of the ARN node, which is a nonnegative number assigned to the graph vertex (see the requirement for the system memory T.7.1: "activation level should be the variable value" [49]).

$l_{\langle tl, v(i), v[j], dir, cnt, r \rangle}$ is the graph edge between the vertices $v[i]$ and $v[j]$, corresponding to the role in the ontology $(i, j \in ID, v[i] = v_{\langle id, tv, brt. \rangle}$

$$tl = TL\left(l_{\langle tl, v(i), v[j], dir, cnt, r \rangle}\right), tl \in TypL,$$

where $TypL = \{type_{l_1}, type_{l_2}, \ldots type_{l_{Nl}}\}$ and $type_{l_i}$ is the type (name) of the edge that uniquely corresponds to the role that unites two specific concepts of the ontology under consideration

$NL = TypL$ – the number of different roles used to build an ontology

$dir = DL(l_{\langle tl, v(i), v[j], dir, cnt, r \rangle}$ is the direction of the edge relative to the vertex, which is indicated first in the edge description (for $l_{\langle tl, v(i), v[j], dir, cnt, r \rangle}$, respectively, to $v[i]$), $dir \in \{in, out\}$; if $dir = in$, then the edge is directed from the vertex specified by the second to the vertex indicated the first and if $dir = out$, then in reverse direction (note: the dir parameter is important because in order to solve practical problems based on the propagation of the resource over the *ARN*, the direction of the edge may affect the result of such propagation)

$cnt = CL(l_{\langle tl, v(i), v[j], dir, cnt, r \rangle})$ is a variable indicating the scope of the concept with the associated *ARN* vertex, from which the edge comes from $cnt \in \{all, some\}$ where all is "everyone," some is "some"

$r = RL(l_{\langle tl, v(i), v[j], dir, cnt, r \rangle})$ is the capacity (conductivity) of the *ARN* edge connecting the vertex $v[i]$ and $v[j]$ (the conductivity of the edge affects the ability to transmit "excitation" from one vertex to another (from one concept to another))

brt is the resource amount, which increases the vertex brightness when accessing it, while the concept brightness in memory should increase in case of activation of any associated concept or in case of its immediate activation [59] (note: an increase in the resource amount (brt), located at the *ARN* apex and uniquely associated with the activated concept, occurs with each invocation of the ontology concept and depends on the invocation type ($init$)).

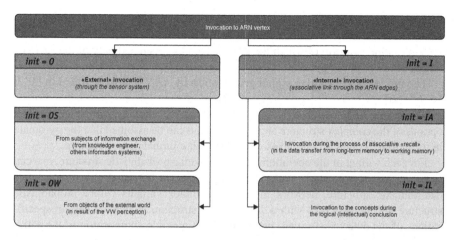

Fig. 4.19 Classification of "excitation" types of the ARN vertices

The classification of the "excitation" variants (*init*) of the *ARN* vertices is shown in Fig. 4.19.

We introduce a function computing the value of the vertex brightness increment upon access to it and depending on the inversion type: $\Delta brt = BRT(init)$, where $init \in \{OS, OW, IA, IL\}$.

It can be assumed that the lowest value should have an increment of Δbrt with $init = IA : BRT(IA)$.

The increments of *BRT (OS)* and *BRT (OW)* in general cases may be equivalent, but it should be noted that the confidence degree of the intelligent system (*IP*) to different subjects of information exchange can be different; therefore, it can be assumed that $BRT (OW) \geq BRT(OS)$. Thus, a peculiar system of priorities for trusting data obtained from various types of sources is declared.

The data transmission on the *ARN* edge leads to an increase in its capacity. It can be reasonably assumed that generally, the value of r, to which the bandwidth of a specific *ARN* edge $l_{\langle tl, v(i), v[j], dir, cnt, r\rangle}$ should be increased when an associative signal between the vertices $v[i]$ and $v[j]$ is transmitted along it, should be proportional to the magnitude of the transmitted signal.

Definition

$\Delta r = THR(BRT(init), TL(l_{\langle tl, v(i), v[j], dir, cnt, r\rangle}), DL(l_{\langle tl, v(i), v[j], dir, cnt, r\rangle}))$ is the value by which the edge capacity increases in case of a signal transmitted over it. Parameters $TL(l_{\langle tl, v(i), v[j], dir, cnt, r\rangle})$ and $DL(l_{\langle tl, v(i), v[j], dir, cnt, r\rangle})$ are entered in the THR ($\bullet$) function in order to more flexibly manage the process of changing the *ARN* edges capacity when transmitting signals over them, if necessary.

Thus, in the functioning, the *IS* must implement the modification of the associative resource network by modifying its structure, the values of its vertex resources, and the conductivity of the edges. All changes in the resource network are carried out only over the activated part of the ontology (i.e., over the part of the ontology and the uniquely corresponding part of the associative resource network). This note allows us to realize the mechanism of a kind of "mapping into memory" of the computing system (on the platform of which the knowledge base operates) only of the "active" fragment of the ontology.

The model of associative signal propagation by *ARN* is demonstrated in [1, 2, 55–62, 66].

Given the *ARN* parameters, uniquely related to ontology, in which various options of the complex structure are represented, it can be assumed that the gyromate in its memory contains complexes of events with a probabilistic structure.

For each variant of the operations' complex with a probabilistic structure, one can construct his own network model (network). Since in this case, any option has a deterministic structure, network models of complex variants with a probabilistic structure will be networks with a deterministic structure, which are certain specifications of the detected (formed) processes. Each possible composition of the activities included in the complex, as well as the links between them, characterize a specific version of the complex structure or, more precisely, a specific feasible implementation of a particular, observable process. A lot of such variants, known to the gyromate and present in its memory in the form of ontology fragments, are not more than countable. Therefore, the procedure for finding and extracting process specifications from the gyromate memory is finite, and if we take into account the possibilities provided by the *AS* expansion mechanism in the *ARN* [1, 2, 8, 19], it is effectively directed.

To make it possible, within the method of destructive impact suppression in heterogeneous network infrastructure using intelligent multi-agent systems, to choose such a strategy of cyber system behavior that would lead to an acceptable exit from a conflict situation developing within the *ITC*, it is necessary for the activities of the simulated process in ontology to be accompanied by temporal parameters that characterize the duration of their implementation.

The developed model of associative resource network [1, 2, 8, 14, 19], working in parallel with ontology, also makes it possible to formalize the mechanism of directional extraction of associative knowledge fragments from the long-term memory of the intelligent system. The proposed model of the associative signal expansion over the *ARN* helps to account for the contexts in extracting and interpreting the knowledge, represented in the memory of the cyber system, and allows us to describe the process of "forgetting" seldom used and false knowledge by reducing their accessibility level. The results obtained in aggregate warrant realizing the procedure of directional knowledge processing by an intelligent system.

The authors' research can be used to develop an intelligent system capable of building in its memory the model of the surrounding cyber environment and synthesizing action programs in accordance with its objectives, which consist in maintaining the level of critically important information infrastructure protection from information and technical impacts, in accordance with this model.

It is safe to assume that the information and technology systems that have the property of anticipation will soon find wide application in various areas of human activities, including the field of ensuring the safety of computer systems belonging to critically important information segments. In the authors' view, the use of anticipatory systems to prevent the destructive effects on network infrastructure objects operating in the *SOPCA* system should increase the level of *CII* protection in time of cyber-attacks.

References

1. Biryukov, D.N.: Cognitive-functional memory specification for simulation of purposeful behavior of cyber systems. Proc. SPIIRAS. **3**(40), 55–76 (2015)
2. Bocharov, V.A., Markin, V.I.: Fundamentals of Logic. Moscow State University, Moscow (2008)
3. Information Operations. Directive TS 3600.1. U. S. Department of Defense, Washington, DC. August 14, 2006 [Electronic resource]. Access mode: https://www.fas.org/irp/doddir/dod/info_ops.pdf
4. Kaspersky, E.: Computer Malignity, 208 p. Peter, St. Petersburg (2008)
5. Levin, I.I., Dordopulo, A.I., Kalyaev, I.A., Doronchenko, Y.I., Razkladkin, M.K.: Modern and promising high-performance computing systems with reconfigurable architecture. Proceedings of the international scientific conference "Parallel Computing Technologies (PaVT'2015)", Ekaterinburg, March 31–April 2, 2015, pp. 188–199. Publishing Center of SUSU, Chelyabinsk (2015)
6. Abramov, S.M.: Research in the field of supercomputer technologies of the IPS RAS: a retrospective and perspective. In: Proceedings of the International Conference "Software Systems: Theory and Applications", vol. 1, pp. 153–192. Publishing house "University of Pereslavl", Pereslavl (2009)
7. Abramov, S.M., Lilitko, E.P.: State and prospects of ultra-high performance computing systems development. Inf. Technol. Comput. Syst. **2**, 6–22 (2013)
8. Petrenko, A.S., Petrenko, S.A.: Super-productive monitoring centers for security threats. Part 1. Protect. Inf. Inside. **2**(74), 29–36 (2017)
9. Petrenko, A.S., Petrenko, S.A.: Designing of corporate segment SOPKA. Protect. Inf. Inside. **6**(72), 48–50 (2016)
10. Petrenko, A.S., Petrenko, S.A.: Super-productive monitoring centers for security threats. Part 2. Protect. Inf. Inside. **3**(75), 48–57 (2017)
11. Petrenko, S.A., Kurbatov, V.A., Bugaev, I.A., Petrenko, A.S.: Cognitive system of early warning about computer attack. Protect. Inf. Inside. **3**(69), 74–82 (2016)
12. Petrenko, S.A., Asadullin, A.Y., Petrenko, A.S.: Evolution of the von Neumann architecture. Protect. Inf. Inside. **2**(74), 18–28 (2017)
13. Klabukov, I.D., Alekhin, M.D., Nekhina, A.A.: The DARPA research program for 2015. Moscow (2014)
14. Petrenko, A.A., Petrenko, S.A.: Research and Development Agency DARPA in the field of cybersecurity. Quest. Cybersecurity. **4**(12), 2–22 (2015)
15. Petrenko, S.A., Petrenko, A.S.: Lecture 12. Perspective tasks of information security. Intelligent information radiophysical systems. Introductory lectures [A. O. Armyakov and others; ed. S.F. Boev, D.D. Stupin, A.A. Kochkarova], pp. 155–166. MSTU them. N.E. Bauman, Moscow (2016)
16. Petrenko, S.A.: The Cyber Threat model on innovation analytics DARPA. Trudy SPII RAN. **39**, 26–41 (2015)
17. Petrov, A.P.: On the perceptron's possibilities. Izvestiya AN SSSR, Technical Cybernetics. 6 (1964)
18. Petrenko, A.S., Bugaev, I.A., Petrenko, S.A.: Master data management system SOPKA. Inf. Protect. Inside. **5**(71), 37–43 (2016)

19. Petrenko, S.A.: Methods of detecting intrusions and anomalies of the functioning of cyber system, Proceedings of ISA RAS. Risk Manag. Safety. **41**, 194–202 (2009)
20. Velichkovsky, B.M.: Cognitive Technical Systems. Computers, Brain, Cognition: Successes of Cognitive Sciences, pp. 273–292. Nauka, Moscow (2008)
21. Petrenko, A.S., Petrenko, S.A.: Large data technologies (BigData) in the field of information security. Inf. Protect. Inside. **4**(70), 82–88 (2016)
22. Petrenko, S.A., Shamsutdinov, T.I., Petrenko, A.S.: Scientific and technical problems of development of situational centers in the Russian Federation. Inf. Protect. Inside. **6**(72), 37–43 (2016)
23. Aristotle. Comp. in 4 volumes (Series "Philosophical heritage"). Thought, Moscow. (1975–1983)
24. Bongard, M.M.: The Problem of Recognition. Fizmatgiz, Moscow (1967)
25. Ryzhikov, Y.I.: Work on the Thesis on Technical Sciences, 496 p. BHV-Petersburg, St. Petersburg (2005)
26. Kolmogorov, A.N.: Automats and life. In: Berg, A.I., Kolman, E. (eds.) Cybernetics: Expected and Cybernetics Unexpected, pp. 12–30. Science, Moscow (1968)
27. Pospelov, D.A.: The modeling of reasoning. Experience in the analysis of mental acts, 184 p. Radio and communication, Moscow (1989)
28. Pospelov, D.A.: Thinking and Automatons, 224 p. 130. Soviet radio, Moscow (1972)
29. Redko, V.G.: Evolution, Neural Networks, Intellect. LIBROKOM Book House/URSS, Moscow (2013)
30. Tarasov, V.B.: System-organizational approach in artificial intelligence. Softw. Prod. Syst. **3**, 6–13 (1999)
31. Marr, B.: Big Data: Using SMART Big Data, Analytics and Metrics to Make Better Decisions and Improve Performance, 246 c. Wiley, New York (2015)
32. Massel, L.V.: Problems of smart grid creation in Russia from the perspective of information technologies and cyber security. In: Proceedings of the All-Russian Seminar with International Participation: Methodological Issues of Research into the Reliability of Large Energy Systems. Vol. 64. Reliability of energy systems: achievements, problems, prospects, pp. 171–181. ISEM SB RAS, Irkutsk (2014)
33. Petrenko, S.A., Petrenko, A.A.: Ontology of cyber-security of self-healing SmartGrid. Protect. Inf. Inside. **2**(68), 12–24 (2016)
34. Petrenko, S.A., Petrenko, A.S.: Practice of application of GOST R IEC 61508. Inf. Protect. Insider. **2**(68), 42–49 (2016)
35. Petrenko, A.A., Petrenko, S.A.: Cyber units: methodical recommendations of ENISA. Quest. Cybersecurity. **3**(11), 2–14 (2015)
36. Petrenko, A.A., Petrenko, S.A.: Intranet Security Audit (Information Technologies for Engineers), 416 p. DMK Press, Moscow (2002)
37. Petrenko, A.A., Petrenko, S.A.: The way to increase the stability of LTE-network in the conditions of destructive cyber-attacks. Quest. Cybersecurity. **2**(10), 36–42 (2015)
38. Petrenko, A.S., Petrenko, S.A.: The first interstate cyber-training of the CIS countries: "Cyber-Antiterror2016". Inf. Protect. Inside. **5**(71), 57–63 (2016)
39. Petrenko, S.A.: Methods of ensuring the stability of the functioning of cyber systems under conditions of destructive effects. Proceedings of the ISA RAS. Risk Manag. Security, **52**, 106–151 (2010)
40. Petrenko, S.A.: Methods of Information and Technical Impact on Cyber Systems and Possible Countermeasures. Proceedings of ISA RAS. Risk Manag. Security, **41**, 104–146 (2009)
41. Petrenko, S.A., Petrenko, A.A.: Creation of a cognitive supercomputer for the computer attacks prevention. Protect Inf. Inside. **3**(75), 14–22 (2017)
42. Petrenko, S.A., Petrenko, A.A.: From detection to prevention: trends and prospects of development of situational centers in the Russian Federation. Intellect Technol. **1**(12), 68–71 (2017)
43. Petrenko, S.A., Petrenko, A.A.: New doctrine as an impulse for the development of domestic information security technologies. Intellect Technol. **2**(13), 70–75 (2017)

44. Petrenko, S.A., Petrenko, A.S.: New doctrine of information security of the Russian Federation. Inf. Protect. Inside. **1**(73), 33–39 (2017)

45. Petrenko, S.A., Simonov, S.V.: Management of Information Risks. Economically Justified Safety (Information technology for engineers), 384 p. DMK-Press, Moscow (2004)

46. Petrenko, S.A.: The concept of maintaining the efficiency of cyber system in the context of information and technical impacts. Proceedings of the ISA RAS. Risk Manag. Safety. **41**, 175–193 (2009)

47. Petrenko, S.A.: The problem of the stability of the functioning of cyber systems under the conditions of destructive effects. Proceedings of the ISA RAS. Risk Manag. Security. **52**, 68–105 (2010)

48. Petrenko, S.A., Kurbatov, V.A.: Information Security Policies (Information Technologies for Engineers), 400 p. DMK Press, Moscow (2005)

49. Abramov, S.M.: History of development and implementation of a series of Russian supercomputers with cluster architecture. In: History of Domestic Electronic Computers. 2nd edn, Rev. and additional; color. Ill.: Publishing house "Capital Encyclopedia", Moscow (2016)

50. Action plan. Document WSIS-03/GENEVA/DOC/5-R dated December 12, 2013. Geneva [Electronic resource]. Access mode: http://www.itu.int/dms_pub/itus/md/03/wsis/doc/S03-WSIS-DOC-0005*PDF-R.pdf

51. Active Engagement, Modern Defence. Strategic Concept for the Defence and Security of the Members of the North Atlantic Treaty Organisation adopted by Heads of State and Government in Lisbon. November 19, 2010 [Electronic resource]. Access mode: http://www.nato.int/cps/en/SID-14EF0623-198FC77E/natolive/official_texts_68580.htm

52. Administration Strategy On Mitigating The Theft Of U.S. Trade Secrets. Executive Office of the President of the United States. February 2013, Washington, DC [Electronic resource]. Access mode: http://www.whitehouse.gov/sites/default/files/omb/IPEC/admin_strategy_on_mitigating_the_theft_of_u.s._trade_secrets.pdf

53. Advances in the field of information and telecommunications in the context of international security. Report of the UN Secretary-General. Document A/66/152 of 15 July 2011 [Electronic resource]. Access mode: http://www.un.org/en/documents/ods.asp?m=A/66/152

54. Advances in the field of information and telecommunications in the context of international security. Report of the First Committee. Document A/66/407 dated November 10, 2011 [Electronic resource]. Access mode: http://www.un.org/en/documents/ods.asp?m=A/66/407

55. Biryukov, D.N., Glukhov, A.P., Pilkevich, S.V., Sabirov, T.R.: Approach to the processing of knowledge in the memory of an intellectual system. Natur. Tech. Sci. **11**, 455–466 (2015)

56. Biryukov, D.N., Lomako, A.G.: Approach to the construction of information security systems capable of synthesizing scenarios of anticipatory behavior in the information conflict. Protect. Inf. Inside. **6**(60), 42–50 (2014)

57. Biryukov, D.N., Lomako, A.G.: Denotational semantics of knowledge contexts in ontological modeling of the subject areas of conflict. Proc. SPIIRAS. **5**(42), 155–179 (2015)

58. Biryukov, D.N., Lomako, A.G.: The formalization of semantics for representation of knowledge about the behavior of conflicting parties: materials of the 22nd scientific-practical conference "Methods and technical means of information security", pp. 8–11. Publishing house of Polytechnic University, St. Petersburg (2013)

59. Biryukov, D.N., Lomako, A.G., Petrenko, S.A.: Generating scenarios for preventing cyber-attacks. Protect. Inf. Inside. **4**(76) (2017)

60. Biryukov, D.N., Rostovtsev, Y.G.: Approach to constructing a consistent theory of synthesis of scenarios of anticipatory behavior in a conflict. Proc. SPIIRAS. **1**(38), 94–111 (2015)

61. Biryukov, D.N., Lomako, A.G., Sabirov, T.R.: Multilevel Modeling of Pre-Emptive Behavior Scenarios. Problems of Information Security. Computer systems, vol. 4, pp. 41–50. Publishing house of Polytechnic University, St. Petersburg (2014)

62. Biryukov, D.N., Lomako, A.G., Rostovtsev, Y.G.: The appearance of anticipatory systems to prevent the risks of cyber threat realization. Proc. SPIIRAS. **2**(39), 5–25 (2015)

63. Scott, D.S.: Models for various type-free calculi. Logic, Methodology and Philosophy of Science IV (Proc. Int. Congress 1971), pp. 157–188. North-Holland (1973)
64. Scott, D.S.: Outline of mathematical theory. 4th Annual Princeton Conference on Information Sciences and Systems, Princeton University, pp. 169–176 (1970)
65. Scott, D.S.: Logic and programming languages. Lectures of the winners of the Turing Award, pp. 65–83; [ed. R. Eschenhurst]. Mir, Moscow (1993)
66. Biryukov, D.N., Lomako, A.G.: Approach to Building a Cyber Threat Prevention System. Problems of Information Security. Computer systems, vol. 2, pp. 13–19. Publishing house of Polytechnic University, St. Petersburg (2013)

Conclusion

Dear reader,

We hope that our book was interesting and beneficial to you!

Indeed, in a time where technologically advanced countries are witnessing an increase in national security threats to informational assets, an early-warning cyber-security system for critically important governmental information assets aimed at ensuring digital sovereignty and defensibility is a particularly relevant concern.

The main information security threats owe to the use of information and communication technologies:

- For hostile purposes in international conflicts, including the disabling of critical infrastructures
- For terrorist exploitation
- For criminal purposes and activities
- As a domination factor to the detriment of the security interests of foreign states

Leading political scientists take a comprehensive view on information and cyberwar, defined as interstate confrontation in the information space aimed at damaging critically important and other information systems, processes, and resources; undermining political, economic, and social systems; brainwashing the population to destabilize society and government; as well as forcing the state to take decisions in the interests of the adverse party.

Earlier, within the framework of international expert cooperation, the EastWest Institute and the M.V. Lomonosov Information Security Institute of Moscow State University, a general conceptual framework for information security was developed. Twenty basic terms were agreed upon, which can be divided into three subgroups:

- Field (cyberspace, cyber infrastructure, cyber service, critical cyberspace, critical cyber infrastructure, critical cyber services)
- Types of threats (cybercrime, cyberterrorism, cyber conflict, cyberwar, cybersecurity)

S. Petrenko, *Big Data Technologies for Monitoring of Computer Security: A Case Study of the Russian Federation*, https://doi.org/10.1007/978-3-319-79036-7

- Actions (fighting in cyberspace, cyber-attack, cyber counterattack, defense and counteraction in cyberspace, cyber war, defensive capabilities in cyberspace, attack potential in cyberspace, the use of benefits in cyberspace, cyber deterrents)

A number of issues devoted to interstate information confrontation in cyberspace were investigated, namely, the applicability of the Hague and Geneva Conventions in compliance with international humanitarian law in the context of war. It was noted that the implementation of the mentioned provisions to cyberspace is determined by the will and capacity of both negotiating parties to conduct a comprehensive analysis of the concepts of cyberwar and cyber weapons, which remain insufficiently defined. Furthermore, corollaries should be drawn for armaments prohibited by the additional 1977 protocols to the Geneva Convention for the Protection of Civilian Persons during war.

In this regard, the development of a national early-warning system for national defense against cyber-attacks on critically important infrastructure is a pertinent scientific-technical problem requiring a solution.

The authors' approach to solving this problem yielded the following results:

1. Ontological model of a structured and context-dependent presentation of knowledge concerning processes of cyberspace conflict on the basis of the Arnold catastrophe theory
2. Functional model of the cognitive early-warning cybersecurity system for critically important governmental information assets
3. Experimental prototypes of software hardware complexes of the cognitive early-warning cybersecurity system based on the theory of multilevel hierarchical systems and methods of "cognitive computations"
4. Common techniques for early-warning cybersecurity system for critically important governmental information assets implemented through the following operations:

   - Input and initial data and knowledge processing used to form the specifications of early-warning cyber-attack in the context of information confrontation
   - Laying down specifications for early-warning cyber-attack operations in information technical conflict based on "cognitive calculations" and logical inference by analogy
   - Data and knowledge preparation and presentation for the prediction of possible outcomes of information conflicts in state cyberspace and selection of process specifications potentially leading to the prevention of cyber-attacks on critical infrastructure as a whole

The scientific novelty of the results, conclusions, and recommendations lies in the fact that it has now become possible to propose specifications for constructing an early-warning cybersecurity system for critically important governmental information assets in the context of information confrontation. The system mentioned can also protect state infrastructure from destructive impact. The scientific and methodological framework of the semantic-syntactic processes modeling was developed, providing the realization of the early-warning scenarios synthesis and proactive behavior. At the same time, implementing a "cognitive" approach in creating this system made it possible for the first time to maintain special knowledge for processing the productivity of information confrontations.

The theoretical significance and scientific value of the obtained results are:

- Formalization of a structured knowledge system of the early-warning cognitive system for cyber-attacks using the complementary formal semantic models: denotational semantics of computable structures, axiomatic semantics of computability properties, and operational semantics of computational actions
- Synthesizing of the intensional expansion of cybersecurity ontologies for an early-warning cybersecurity system for constructing an arbitrarily complex multi-tilayered ontological construction
- Development of a consistent intensional theory of partially ordered structured models of functioning for formalizing the semantics of context-dependent knowledge concerning the conflict subject
- Construction of a formal axiomatized solvable interpreted theory for modeling the data and context-dependent knowledge hierarchical representation and generation of truth-based reasoning, applying the semantics of roles in the mentioned multilayered cybersecurity ontologies
- Construction of a hierarchical level-by-level coordinated early-warning system for cyber-attacks, based on the theory of hierarchical systems and methods of "cognitive calculations" used to search for potential solution algorithms
- Proof of model completeness, consistency, and the named theory solvability

Practical significance of the obtained results is the outcome of work that has been brought to a level suitable for practical testing and used for:

- Design and development of prototypes and an open segment range of the national early-warning cybersecurity system for critically important government information assets, ensuring a general increase in the completeness of the parameters monitored
- Construction of formalized models for cybersecurity threat, providing increased and detailed depth of the forecasts for multilayered potentially implemented attacks on critically important state information assets
- Generation of process specifications to be potentially implemented in information conflict
- Development of private methods for giving early warning of and preventing potential attacks
- Setting up and conducting national and international cyber-lessons, as well as various types of training sessions, seminars, and practices aimed at improving and developing models and methods for an early-warning cybersecurity system for critically important governmental information assets

In conclusion, we would like to note that the theory and practice of early-warning cybersecurity system for critically important state information assets are just emerging and thus are in the initial stage of their formation and development. Therefore, our monograph is primarily a first attempt to share with beginners and experienced specialists the experience available in research and development in the field as well as to develop some practical recommendations. We intend to continue our work in this direction, and for this your feedback, dear readers, is required. Taking this into account, we will be very grateful for any comments and suggestions on expanding and improving the quality of the book's material. You may contact the authors at these email addresses:

s.petrenko@rambler.ru and a.petrenko1999@rambler.ru

Russia-Germany
January 2018

Definition List

AC&S	Access Control and Security
ACL	Access control list
AuthN/AuthZ	Authentication/authorization
AP	Access point
APT	Advanced persistent threat
ARP	Address Resolution Protocol
CAIDA	Cooperative Association for Internet Data Analysis
CDC US	Centers for Disease Control and Prevention
CEP	Complex event processing
CERT	Computer emergency response team
CIA	Confidentiality, integrity, and availability
CIAC	Computer Incident Advisory Capability
CISO	Chief information security officers
CLI	Command-line interface
CMC	Crisis management center
CMVP	Cryptographic Module Validation Program
CNCI	Comprehensive National Cybersecurity Initiative
COM	Component Object Model
CPU	Central processing unit
CSA	Cloud Security Alliance
CSA BDWG	Cloud Security Alliance Big Data Working Group
CSIRT	Computer security incident response team
CSIS	Center for Strategic and International Studies
CSP	Cloud service provider
CSRC	Computer Security Resource Center
CSV	Comma-separated values
CVE	Common Vulnerabilities and Exposures
CNCI	Comprehensive National Cybersecurity Initiative
CTF	"Capture the flag" game

CSIRT	Computer Security Incident Response Team
DARPA	Defense Advanced Research Projects Agency
DDoS	Distributed denial-of-service
DHCP	Dynamic Host Configuration Protocol
DISA	Defense Information Systems Agency
DLL	Dynamic-link library
DMZ	Demilitarized zone
DNS	Domain Name System
DOD US	Department of Defense
DoS	Denial of service
DS	Distribution system
DShield	Distributed Intrusion Detection System
EICAR	European Institute for Computer Antivirus Research
ESP	Encapsulating Security Payload
EU	European Union
FIPS	Federal Information Processing Standards
FISMA	Federal Information Security Management Act
FPLG	Field programmable logic devices
FSTEC	Federal Service for Technical and Export Control
FTP	File Transfer Protocol
GIG	Global Information Grid
GHz	Gigahertz
GPS	Global Positioning System
GRC	Governance, risk management, and compliance
GUI	Graphical user interface
JSOC	Joint Special Operations Command
JTF-GNO	Joint Tactical Force for Global Network Operations
JFCCNW	Joint Functional Component Command for Network Warfare
HAP	High Assurance Platform
HSM	Hardware Security Module technology
HTTP	Hypertext Transfer Protocol
HTTPS	Hypertext Transfer Protocol over SSL
ICT	Information and computer technology
ICMP	Internet Control Message Protocol
IDPS	Intrusion detection and prevention system
IDS	Intrusion detection system
IEEE	Institute of Electrical and Electronics Engineers
IETF	Internet Engineering Task Force
IGMP	Internet Group Management Protocol
IM	Instant messaging
IMAP	Internet Message Access Protocol
IP	Internet Protocol
IPIB	Institute of Information Security Problems
IPS	Intrusion prevention system

IPsec	Internet Protocol Security
IRC	Internet Relay Chat
ISC	Internet Storm Center
IoT	Internet of Things
IT	Information technology
ITL	Information Technology Laboratory
LAN	Local area network
MAC	Media access control
MDR	Managed Detection and Response Services
MIFT	Moscow Institute of Physics and Technology
MSSP	Managed security service provider
M2M	Machine to machine
NBA	Network Behavior Analysis
NCSD	National Cyber Security Division
NBAD	Network behavior anomaly detection
NFAT	Network Forensic Analysis Tool
NFS	Network File System
NFV	Network functions virtualization
NIC	Network interface card
NIEM	National Information Exchange Model
NIST	National Institute of Standards and Technology
NSA	National Security Agency
NTP	Network Time Protocol
NVD	National Vulnerability Database
OMB	Office of Management and Budget
OS	Operating system
OSS	Operations systems support
PaaS	Platform as a Service
PKI	Public key infrastructure
PDA	Personal digital assistant
PoE	Power over Ethernet
POP	Post Office Protocol
RF	Radio frequency
RFC	Request for Comment
ROM	Read-only memory
RPC	Remote procedure call
SAML	Security Assertion Markup Language
SATOSA	System of traffic analysis and network attack detection
SDN	Software-defined networking technology
SEM	Security event management
SIEM	Security information and event management
SIM	Security information management
SIP	Session Initiation Protocol
SLA	Service-level agreement

SMB	Server Message Block
SMTP	Simple Mail Transfer Protocol
SNMP	Simple Network Management Protocol
SOC	Security operations center
SOPCA	Prevention and Cyber Security Incident Response
SPOCA	Computer attack detection and prevention system
SP	Special Publication
SSH	Secure Shell
SSID	Service set identifier
SSL	Secure Sockets Layer
STA	Station
STS	Security Token Service
TCP	Transmission Control Protocol
TCP/IP	Transmission Control Protocol/Internet Protocol
TFTP	Trivial File Transfer Protocol
TLS	Transport Layer Security
TTL	Time to live
USE	Unified State Examination
UDP	User Datagram Protocol
USB	Universal Serial Bus
US-CERT	United States Computer Emergency Readiness Team
VLAN	Virtual Local Area Network
VM	Virtual machine
VPN	Virtual private network
WEP	Wired Equivalent Privacy
WLAN	Wireless local area network
WPA	Wi-Fi Protected Access
WVE	Wireless Vulnerabilities and Exploits
XACML	eXtensible Access Control Markup Language
XML	Extensible Markup Language

Glossary

Agent A host-based intrusion detection and prevention program that monitors and analyzes activity and may also perform prevention actions.

Alert A notification of an important observed event.

Anomaly-Based Detection The process of comparing definitions of what activity is considered normal against observed events to identify significant deviations.

Antivirus Software A program that monitors a computer or network to identify all major types of malware and prevent or contain malware incidents.

Application-Based Intrusion Detection and Prevention System A host-based intrusion detection and prevention system that performs monitoring for a specific application service only, such as a Web server program or a database server program.

Blacklist A list of discrete entities, such as hosts or applications, that have been previously determined to be associated with malicious activity.

Blinding Generating network traffic that is likely to trigger many alerts in a short period of time, to conceal alerts triggered by a "real" attack performed simultaneously.

Channel Scanning Changing the channel being monitored by a wireless intrusion detection and prevention system.

Console A program that provides user and administrator interfaces to an intrusion detection and prevention system.

Database Server A repository for event information recorded by sensors, agents, or management servers.

Evasion Modifying the format or timing of malicious activity so that its appearance changes but its effect on the target is the same.

False Negative An instance in which an intrusion detection and prevention technology fails to identify malicious activity as being such.

False Positive An instance in which an intrusion detection and prevention technology incorrectly identifies benign activity as being malicious.

Flooding Sending large numbers of messages to a host or network at a high rate. In this publication, it specifically refers to wireless access points.

© Springer International Publishing AG, part of Springer Nature 2018
S. Petrenko, *Big Data Technologies for Monitoring of Computer Security: A Case Study of the Russian Federation*, https://doi.org/10.1007/978-3-319-79036-7

Flow A particular network communication session occurring between hosts.

Host-Based Intrusion Detection and Prevention System A program that monitors the characteristics of a single host and the events occurring within that host to identify and stop suspicious activity.

Incident A violation or imminent threat of violation of computer security policies, acceptable use policies, or standard security practices.

Inline Sensor A sensor deployed so that the network traffic it is monitoring must pass through it.

Intrusion Detection The process of monitoring the events occurring in a computer system or network and analyzing them for signs of possible incidents.

Intrusion Detection and Prevention The process of monitoring the events occurring in a computer system or network, analyzing them for signs of possible incidents, and attempting to stop detected possible incidents. See also "intrusion prevention."

Intrusion Detection System Load Balancer A device that aggregates and directs network traffic to monitoring systems, such as intrusion detection and prevention sensors.

Intrusion Detection System Software that automates the intrusion detection process.

Intrusion Prevention The process of monitoring the events occurring in a computer system or network, analyzing them for signs of possible incidents, and attempting to stop detected possible incidents. See also "intrusion detection and prevention."

Intrusion Prevention System Software that has all the capabilities of an intrusion detection system and can also attempt to stop possible incidents. Also called an intrusion detection and prevention system.

Jamming Emitting electromagnetic energy on a wireless network's frequencies to make them unusable by the network.

Malware A program that is inserted into a system, usually covertly, with the intent of compromising the confidentiality, integrity, or availability of the victim's data, applications, or operating system or of otherwise annoying or disrupting the victim.

Management Network A separate network strictly designed for security software management.

Management Server A centralized device that receives information from sensors or agents and manages them.

Network-Based Intrusion Detection and Prevention System An intrusion detection and prevention system that monitors network traffic for particular network segments or devices and analyzes the network and application protocol activity to identify and stop suspicious activity.

Network Behavior Analysis System An intrusion detection and prevention system that examines network traffic to identify and stop threats that generate unusual traffic flows.

Network Tap A direct connection between a sensor and the physical network media itself, such as a fiber optic cable.

Passive Fingerprinting Analyzing packet headers for certain unusual characteristics or combinations of characteristics that are exhibited by particular operating systems or applications.

Passive Sensor A sensor that is deployed so that it monitors a copy of the actual network traffic.

Promiscuous Mode A configuration setting for a network interface card that causes it to accept all incoming packets that it sees, regardless of their intended destinations.

Sensor An intrusion detection and prevention system component that monitors and analyzes network activity and may also perform prevention actions.

Shim A layer of host-based intrusion detection and prevention code placed between existing layers of code on a host that intercepts data and analyzes it.

Signature A pattern that corresponds to a known threat.

Signature-Based Detection The process of comparing signatures against observed events to identify possible incidents.

Spanning Port A switch port that can see all network traffic going through the switch.

Stateful Protocol Analysis The process of comparing predetermined profiles of generally accepted definitions of benign protocol activity for each protocol state against observed events to identify deviations.

Stealth Mode Operating an intrusion detection and prevention sensor without IP addresses assigned to its monitoring network interfaces.

Threshold A value that sets the limit between normal and abnormal behavior.

Triangulation Identifying the physical location of a detected threat against a wireless network by estimating the threat's approximate distance from multiple wireless sensors by the strength of the threat's signal received by each sensor and then calculating the physical location at which the threat would be the estimated distance from each sensor.

Tuning Altering the configuration of an intrusion detection and prevention system to improve its detection accuracy.

Whitelist A list of discrete entities, such as hosts or applications, that are known to be benign.

Wireless Intrusion Detection and Prevention System An intrusion detection and prevention system that monitors wireless network traffic and analyzes its wireless networking protocols to identify and stop suspicious activity involving the protocols themselves.

References

1. About formal bases of OWL [Electronic resource]. Access mode: http://semanticfuture.net/index.php. Accessed 20 Dec 2014
2. Abramov, S.M.: Research in the field of supercomputer technologies of the IPS RAS: a retrospective and perspective. In: Proceedings of the International Conference "Software Systems: Theory and Applications", vol. 1, pp. 153–192. Publishing house "University of Pereslavl", Pereslavl (2009)
3. Abramov, S.M.: History of development and implementation of a series of Russian supercomputers with cluster architecture. In: History of Domestic Electronic Computers. 2nd edn, Rev. and additional; color. Ill.: Publishing house "Capital Encyclopedia", Moscow (2016)
4. Abramov, S.M., Lilitko, E.P.: State and prospects of ultra-high performance computing systems development. Inf. Technol. Comput. Syst. **2**, 6–22 (2013)
5. Action plan. Document WSIS-03/GENEVA/DOC/5-R dated December 12, 2013. Geneva [Electronic resource]. Access mode: http://www.itu.int/dms_pub/itus/md/03/wsis/doc/S03-WSIS-DOC-0005*PDF-R.pdf
6. Active Engagement, Modern Defence. Strategic Concept for the Defence and Security of the Members of the North Atlantic Treaty Organisation adopted by Heads of State and Government in Lisbon. November 19, 2010 [Electronic resource]. Access mode: http://www.nato.int/cps/en/SID-14EF0623-198FC77E/natolive/official_texts_68580.htm
7. Administration Strategy On Mitigating The Theft Of U.S. Trade Secrets. Executive Office of the President of the United States. February 2013, Washington, DC [Electronic resource]. Access mode: http://www.whitehouse.gov/sites/default/files/omb/IPEC/admin_strategy_on_mitigating_the_theft_of_u.s._trade_secrets.pdf
8. Advances in the field of information and telecommunications in the context of international security. Report of the UN Secretary-General. Document A/66/152 of 15 July 2011 [Electronic resource]. Access mode: http://www.un.org/en/documents/ods.asp?m=A/66/152
9. Advances in the field of information and telecommunications in the context of international security. Report of the First Committee. Document A/66/407 dated November 10, 2011 [Electronic resource]. Access mode: http://www.un.org/en/documents/ods.asp?m=A/66/407
10. Advances in the field of information and telecommunications in the context of international security. Resolution of the General Assembly of the UN. Document A/RES/65/41 dated December 8, 2010 [Electronic resource]. Access mode: http://www.un.org/en/documents/ods.asp?m=A/RES/65/41
11. Advances in the field of information and telecommunications in the context of international security. Resolution of the General Assembly of the UN. Document A/RES/68/243 dated

© Springer International Publishing AG, part of Springer Nature 2018
S. Petrenko, *Big Data Technologies for Monitoring of Computer Security: A Case Study of the Russian Federation*, https://doi.org/10.1007/978-3-319-79036-7

December 27, 2013 [Electronic resource]. Access mode: http://www.un.org/en/ga/search/view_doc.asp?symbol=A/RES/68/243

12. Advancing America's Networking and Information Technology Research and Development Act of 2013. H. R. 967 [Electronic resource]. Access mode: https://www.govtrack.us/congress/bills/113/hr967/text

13. Agreement between the governments of the member states of the Shanghai Cooperation Organization on cooperation in the field of international information security from June 16, 2009, Yekaterinburg. Appendix 1. [Electronic resource]. Access mode: https://ccdcoe.org/sites/default/files/documents/SCO-090616-IISAgreementRussian.pdf

14. Aldrich, R.W.: The International Legal Implications of Information Warfare [Electronic resource]. Airpower J. **10**(3), 99–110 (1996). Access mode: http://www.airpower.maxwell.af.mil/airchronicles/apj/apj96/fall96/aldrich.pdf

15. Alekseeva, I.Y., et al.: Information Challenges of National and International Security; [under the Society. ed. A. V. Fedorova, VN Tsigichko], 328 p. PIR Center, Moscow (2001)

16. Alessandri, D., et al.: Towards a Taxonomy of Intrusion-Detection Systems and Attacks. Zurich, IBM Research Division (2001)

17. Almgren, M.: Consolidation and evaluation of IDS taxonomies. In: Proceedings of the Eight Nordic Workshop on Secure IT Systems, NordSec 2003

18. An evaluation Framework for National Cyber Security Strategies [Electronic resource]. European Union Agency for Network and Information Security (2014). Access mode: https://www.enisa.europa.eu/activities/Resilienceand-CIIP/national-cyber-security-strategies-ncsss/an-evaluation-framework-for-cyber-security-strategies-1

19. An Open, Safe and Secure Cyberspace. Joint communication to the European Parliament, the Council, the European Economic and Social committee and the Committee of the Regions Cybersecurity Strategy of the European Union of the European Commission and Higher Representative for foreign affairs and security policy. Brussels (2013) [Electronic resource]. Access mode: http://ec.europa.eu/information_society/newsroom/

20. Anderson, J.P.: Computer Security Threat Monitoring and Surveillance. James P. Anderson Co., Fort Washington, PA (1980)

21. Andreev, V.V., Zdiruk, K.B.: IV Jupiter: implementation of corporate security policy in computer networks. Open. Syst. **7–8**, 43–46 (2003)

22. Annual Incident Reports 2014: Analysis of Article 13a annual incident reports / European Union Agency for Network and Information Security (ENISA) (2015). [Electronic resource]. Access mode: https://www.enisa.europa.eu/activities/Resilience-and-CIIP/Incidents-reporting/annual-reports/annual-incident-reports-2014. Accessed 10 Apr 2016

23. Appliance of information and communication technologies for development. Resolution of the General Assembly of the UN. Document A/RES/65/141 dated December 20, 2010 [Electronic resource]. Access mode: http://www.un.org/en/ga/search/view_doc.asp?symbol=A/RES/65/141

24. Arbatov A.G. Real and imaginary threats: Military power in world politics in the beginning of the XXI century. [Electronic resource] AG Arbatov. Russia in global politics. March 3, 2013. Access mode: http://www.global- affairs.ru/number/Ugrozy-realnye-i-mnimye-15863

25. Aristotle. Comp. in 4 volumes (Series "Philosophical heritage"). Thought, Moscow. (1975–1983)

26. Arquilla, J.: Ethics and information warfare. In: Khalilzad, Z., White, J., Marsall, A. (eds.) Strategic Appraisal: The Changing Role of Information in Warfare, 475 p. RAND Corporation, Santa Monica (1999)

27. Ashby, U.R.: Principles of Self-Organization, pp. 314–343. Mir, Moscow (1966)

28. Axelsson, S.: Intrusion Detection Systems: A Taxonomy and Survey. Technical Report 99–15. Dept of Computer Engineering, Chalmers University of Technology, Goteborg (2000)

29. Barabanov, A.V., Markov, A.S., Tsirlov, V.L.: Methodological framework for analysis and synthesis of a set of secure software development controls. J. Theor. Appl. Info. Technol. **88**(1), 77–88 (2016)

30. Barabanov, A., Lavrov, A., Markov, A., Polotnyanschikov, I., Tsirlov, V.: The study into cross-site request forgery attacks within the framework of analysis of software vulnerabilities. In: Preliminary proceedings of the 11th Spring/Summer Young Researchers' Colloquium on Software Engineering (Innopolis, Republic of Tatarstan, Russian Federation, June 5–7, 2017), pp. 105–109. SYRCoSE, ISP RAS

31. Baranov, P.A.: Detection of anomalies based on the application of the criterion of the dispersion degree. Proceedings of the XIV All-Russian Scientific Conference "Information Security Problems in the Higher School System", pp. 25–27. Izd. department of the St. Petersburg State Polytechnic University, St. Petersburg (2007)

32. Batueva, E.V.: American concept of threats to information security and its international political component, 207 p. Doctoral thesis of political sciences. MGIMO (U) Ministry of Foreign Affairs of the Russian Federation, Moscow (2014)

33. Bedritsky, A.V.: American policy of cyber space control. Probl. Natl. Strat. **2**(3), 25–40 (2010)

34. Bedritsky, A.V.: Information War: Concepts and Their Implementation in the US, 183p. RISI, Moscow (2008)

35. Bedritsky, A.V.: The Evolution of the American Concept of Information War, 26p. RISI, Moscow. Analytical Rev. (3) (2003)

36. Biryukov, D.N.: Cognitive-functional memory specification for simulation of purposeful behavior of cyber systems. Proc. SPIIRAS. **3**(40), 55–76 (2015)

37. Biryukov, D.N., Lomako, A.G.: Denotational semantics of knowledge contexts in ontological modeling of the subject areas of conflict. Proc. SPIIRAS. **5**(42), 155–179 (2015)

38. Biryukov, D.N., Glukhov, A.P., Pilkevich, S.V., Sabirov, T.R.: Approach to the processing of knowledge in the memory of an intellectual system. Natur. Tech. Sci. **11**, 455–466 (2015)

39. Biryukov, D.N., Lomako, A.G.: Approach to the construction of information security systems capable of synthesizing scenarios of anticipatory behavior in the information conflict. Protect. Inf. Inside. **6**(60), 42–50 (2014)

40. Biryukov, D.N., Lomako, A.G.: The formalization of semantics for representation of knowledge about the behavior of conflicting parties: materials of the 22nd scientific-practical conference "Methods and technical means of information security", pp. 8–11. Publishing house of Polytechnic University, St. Petersburg (2013)

41. Biryukov, D.N., Lomako, A.G., Petrenko, S.A.: Generating scenarios for preventing cyber-attacks. Protect. Inf. Inside. **4**(76) (2017)

42. Biryukov, D.N., Lomako, A.G., Rostovtsev, Y.G.: The appearance of anticipatory systems to prevent the risks of cyber threat realization. Proc. SPIIRAS. **2**(39), 5–25 (2015)

43. Biryukov, D.N., Lomako, A.G., Sabirov, T.R.: Multilevel Modeling of Pre-Emptive Behavior Scenarios. Problems of Information Security. Computer systems, vol. 4, pp. 41–50. Publishing house of Polytechnic University, St. Petersburg (2014)

44. Biryukov, D.N., Rostovtsev, Y.G.: Approach to constructing a consistent theory of synthesis of scenarios of anticipatory behavior in a conflict. Proc. SPIIRAS. **1**(38), 94–111 (2015)

45. Biryukov, D.N., Lomako, A.G.: Approach to Building a Cyber Threat Prevention System. Problems of Information Security. Computer systems, vol. 2, pp. 13–19. Publishing house of Polytechnic University, St. Petersburg (2013)

46. Bocharov, V.A., Markin, V.I.: Fundamentals of Logic. Moscow State University, Moscow (2008)

47. Boev, S.F., Kochkarov, A.A., Stupin, D.D.: Development of R & D activities of high-tech B2G-holdings: problems and tasks. Qual. Innov. Educ. **11**(78), 54–59 (2011)

48. Boev, S.F., Kochkarov, A.A., Stupin, D.D.: The role and possibilities of pre-university training in the problem of the formation of highly qualified specialists for high-tech branches of the real economy and the experience of the RTI Systems Concern: materials of the International Scientific Conference "Forming the Identity of Finno-Ugric world and Russian education", pp. 330–333. Mordovian state publishing house University, Saransk (2011)

49. Bongard, M.M.: The Problem of Recognition. Fizmatgiz, Moscow (1967)

50. Brennen, S.: Cyberthreats and the Decline of the Nation-state, 175 p. Susan W. Brenner. Routledge, Abingdon (2014)
51. Brenner, J.: America the Vulnerable, 308 p. Joel Brenner. Penguin Press, New York (2011)
52. Carr, J.: Inside Cyber Warfare, 213 p. Jeffrey Carr. O'Reilly (2010)
53. Cavelty, M.: Cyber-Security and Threat Politics: US Efforts to Secure the Information Age, 182 p. Myriam Dunn Cavelty. Routledge, New York (2007)
54. Chereshkin, D.S.: Problems of Information Security Management, 224 p. Editorial URSS, Moscow (2002)
55. Clarifying Cybersecurity Responsibilities and Activities of the Executive Office of the President and the Department of Homeland Security. Memorandum. Executive Office of the President Office of Management and Budget, Washington, DC. July 6, 2010 [Electronic resource]. Access mode: http://www.whitehouse.gov/sites/default/files/omb/assets/memo-randa_2010/m10-28.pdf
56. Clark, R., Nake, R.: The Third World War. What Will It Be Like? Publishing house "Peter", St. Petersburg (2011)
57. Clark, W., Levin, P.: Securing the information highway: How to enhance the United States electronic defenses. Foreign Aff. November/December 2009 [Electronic resource]. Access mode: http://www.foreignaffairs.com/articles/65499/wesley-k-clark-and-peter-llevin/secur ing-the-information-highway
58. Clarke, R.: Cyber War the Next Threat to National Security and What to Do About It, 290 p. (Richard A. Clarke and Robert K. Knake). HarperCollins (2010)
59. Clarke, R.: Securing Cyberspace Through International Norms. Good Harbor Security Risk Management [Electronic resource]. Access mode: http://www.goodharbor.net/media/pdfs/ SecuringCyberspace_web.pdf
60. Clayton, M.: Presidential Cyberwar Directive Gives Pentagon Long-awaited Marching Orders. The Christian Science Monitor. June 10, 2013 – [Electronic resource]. Access mode: http://www.csmonitor.com/USA/Military/2013/0610/Presidential-cyberwardirective-gives-Pentagon-long-awaited-marching-orders-video
61. Collin, B.: The Future of Cyberterrorism. Crime Justice Int. **13**(2) March 1997 [Electronic resource]. Access mode: http://www.cjimagazine.com/archives/cji4c18.html?id=415
62. Collins, A.M., Quillian, M.R.: Retrieval time from semantic memory. J. Verbal Learn. Verbal Behav. **8**, 240–247 (1969)
63. Communication from the Commission to the European Parliament and the Council. The EU Internal Security Strategy in Action: Five steps towards a more secure Europe. Brussels, 22.11.2010. COM (2010)
64. Comprehensive National Cybersecurity Initiative. The White House, Washington, DC. January 2008 [Electronic resource]. Access mode: http://www.whitehouse.gov/cybersecurity/compre hensive-nationalcybersecurity-initiative
65. Consolidated and Further Continuing Appropriations Act of 2013. H. R. 933 [Electronic resource]. Access mode: http://www.gpo.gov/fdsys/pkg/BILLS-113hr933pp/pdf/BILLS-113hr933pp.pdf
66. Cornish, P.: Cyber security and politically, socially and religiously motivated cyber-attacks. 2009 [Electronic resource]. Access mode: http://www.europarl.europa.eu/activities/commit tees/studies.do?language=EN
67. Creation of a global culture of cybersecurity and assess national efforts to protect critical information infrastructures. UN Resolution. Document A/RES/64/211 dated December 21, 2009 [Electronic resource]. Access mode: http://www.un.org/en/documents/ods.asp? m=A/RES/64/211
68. Crimes involving the use of a computer network. The Tenth United Nations Congress on the Prevention of Crime and the Treatment of Offenders. Document A / CONF.187 / 10 of 3 February 1999
69. Critical Infrastructure Research and Development Advancement Act of 2013. H. R. 2952 [Electronic resource]. Access mode: https://www.govtrack.us/congress/bills/113/hr2952/text

70. Critical Infrastructure Security and Resilience: Presidential Policy Directive/PPD-21. The White House, Washington, DC. February 12, 2013
71. Cyber Europe 2012: Key Findings Report. ENISA. 2012 [Electronic resource]. Access mode: https://www.enisa.europa.eu/activities/Resilience-and-CIIP/cyber-crisis-cooperation/cce/cyber-europe/cyber-europe-2012/cyber-europe2012-key-findings-report.   Accessed   date 10 Apr 2016
72. Cyber Intelligence Sharing and Protection Act. 2012. H. R. 3523 [Electronic resource]. Access mode: https://www.govtrack.us/congress/bills/112/hr3523
73. Cyber Security Report. European Commission. 2013 [Electronic resource]. Access mode: http://ec.europa.eu/public_opinion/archives/ebs/ebs_404_en.pdf. Accessed date 10 Apr 2016
74. Cyber Security Report. European Commission. 2015. [Electronic resource]. Access mode: http://ec.europa.eu/COMMFrontOffice/PublicOpinion/index.cfm/Survey/getSurveyDetail/yearFrom/1973/yearTo/2016/search/cyber/surveyKy/2019. Accessed 10 Apr 2016
75. Cyberpower and National Security [ed. F. Kramer, S. Starr, and L. Wentz], 664 p. Potomac Books Inc. (2009)
76. Cybersecurity Act of 2009. S.773. Open Congress Summary [Electronic resource]. Access mode: http://www.opencongress.org/bill/111-s773/show
77. Cybersecurity Strategy of the European Union: An Open, Safe and Secure Cyberspace. High Representative of the European Union for Foreign Affairs and Security Policy. Brussel, 2013 [Electronic resource]. Access mode: http://eeas.europa.eu/policies/eu-cyber-security/cybsec_comm_en.pdf. Accessed 10 Apr 2016
78. Cyberspace Policy Review Assuring a Trusted and Resilient Information and Communications Infrastructure. May 2009 [Electronic resource]. Access mode: http://www.whitehouse.gov/assets/documents/Cyberspace_Policy_Review_final.pdf
79. Debar H., et al.: (IBM Zurich). Towards a Taxonomy of Intrusion-Detection Systems. IBM Research Division, Zurich (1999)
80. Decree of the Government of the Russian Federation of 04 September 2003 No. 547 "On the preparation of the population in the field of protection from natural and man-made emergency situations"
81. Decree of the Government of the Russian Federation of December 30, 2003 No. 794 "On Unified State System for the Prevention and Elimination of Emergency Situations"
82. Decree of the Government of the Russian Federation of December 8, 2011 No. 2227-r "On the Approval of the Strategy for Innovative Development of the Russian Federation for the Period to 2020". [Electronic resource]. Access mode: http://mon.gov.ru/files/materials/4432/11.12.08-2227r.pdf. 145. RD 50-34.698-90. Automated systems. Requirements for the content of documents
83. Denning, D.: Cyberterrorism. George Town University. May 23, 2000 [Electronic resource]. Access mode: http://www.cs.georgetown.edu/~denning/infosec/cyberterror.html
84. Denning, D.: Information Operations and Terrorism / Defense Technical Information Center. August 18, 2005. [Electronic resource]. Access mode: http://www.dtic.mil/cgi-bin/GetTRDoc?AD=ADA484999
85. Denning, D.: Information Warfare and Security, 522 p. ACM Press, New York (1999)
86. Denning, D.: Is cyberterror next? Social Science Research Council. November 1, 2001 [Electronic resource]. Access mode: http://essays.ssrc.org/sept11/essays/denning.htm
87. Denning, D.: Reflections on cyberweapons controls. Comput. Security J. **XVI**(4), 43–53 (2000)
88. Denning, D.E., (SRI International): An Intrusion Detection Model. IEEE Transactions on Software Engineering (SE-13), **2**, 222–232 (1987)
89. Department of Defense Dictionary of Military and Associated Terms. November 8, 2010 [Electronic resource]. Access mode: http://www.dtic.mil/doctrine/dod_dictionary/
90. Department of Defense Strategy for Operating in Cyberspace. July 2011. [Electronic resource]. Access mode: http://www.defense.gov/news/d20110714cyber.pdf

91. Digital Agenda for Europe. A Europe 2020 Strategy. 2010 [Electronic resource]. Access mode: http://ec.europa.eu/digitalagenda.
92. Dunlap, C. Jr.: Perspectives for cyber strategists on law for cyberwar (Charles J. Dunlap Jr.). Strateg. Stud. Q. Spring, 81–99 (2011)
93. Electronic Communications Privacy Act Amendments Act of 2013. S. 607 [Electronic resource]. Access mode: https://www.govtrack.us/congress/bills/113/s607
94. Elliott, D.: Weighing the Case for a Convention to Limit Cyberwarfare. Arms Control Association. November 2009 [Electronic resource]. Access mode: http://www.armscontrol. org/act/2009_11/Elliott
95. Ermakov, S.M.: Transformation of NATO after the Lisbon Summit in 2010: from the defense of the territory to the protection of the public domain. Probl. Natl. Strateg. 4(9), 107–128 (2011)
96. Establishing the Office of Homeland Security and the Homeland Security Council: Executive Order 13228. The White House, Washington, DC. October 8, 2001. [Electronic resource]. Access mode: http://www.fas.org/irp/offdocs/eo/eo-13228.htm
97. Exaflop technology. The concept on the development of high-performance computing technology on the basis of superframe exaflop class (2012–2020), 111 p. State Corporation "Rosatom" and others, Moscow (2015)
98. Expressing the sense of Congress regarding actions to preserve and advance the multistakeholder governance model under which the Internet has thrived: Congress Resolution. S. CON.RES.50. 112th Congress. June 27, 2012. [Electronic resource]. Access mode: http:// www.gpo.gov/fdsys/pkg/BILLS-112sconres50is.pdf
99. Federal Information Security Amendments Act of 2013. H. R. 1163 [Electronic resource]. Access mode: http://beta.congress.gov/bill/113thcongress/house-bill/1163
100. Federal Information Security Management Act of 2002. Title III of the EGovernment Act of 2002, Public Law 107–347, 44 U.S.C. 3541 [Electronic resource]. Access mode: http://csrc. nist.gov/drivers/documents/FISMA-final.pdf
101. Federal Law No. 149-FZ of July 27, 2006 (edition of July 6, 2016) "On Information, Information Technologies and Information Protection"
102. Federal Law of 06 March 2006 No. 35-FZ "On Countering Terrorism"
103. Federal Law of the Russian Federation of December 21, 1994 No. 68-FZ "On Protection of the Population and Territories from Emergencies of Natural and Technogenic Character"
104. Federal Law of the Russian Federation of July 27, 2006 No. 152-FZ "On Personal Data" (edition of February 22, 2017)
105. Federal Service for Technical and Export Control Order No. 31 of March 14, 2014 "On Approving the Requirements for Providing Information Protection in Automated Control Systems of Production and Technological Processes at Critical Facilities, Potentially Hazardous Facilities, and Objects of Increased Danger to Life and Health of People and for the environment"
106. Federal Service for Technical and Export Control Order of Russia of February 18, 2013 No. 21 "On the approval of the composition and content of organizational and technical measures to ensure the safety of personal data when processing them in personal data information systems"
107. Federal Service for Technical and Export Control Order of Russia of 11 February 2013 No. 17 "On approval of the requirements for the protection of information that is not classified as a state secret contained in government information systems"
108. Finland's Cyber security Strategy. Forssa print, Finland, 2013 [Electronic resource]. Access mode: https://www.enisa.europa.eu/activities/Resilience-and-CIIP/national-cyber-securitystrategies-ncsss/FinlandsCyberSecurityStrategy.pdf. Accessed 10 Apr 2017
109. Finn, V.K.: Artificial Intelligence: The Ideological Base and the Main Product. In: Proceedings of the 9th National Conference on Artificial Intelligence, vol. 1, pp. 11–20. Fizmatlit, Moscow (2004)
110. Finn, V.K.: On the intellectual analysis of data. News. Artificial Intel. 3, 3–18 (2004)

111. Friedman, T.: The Lexus and the Olive Tree: Understanding Globalization, 394 p. Thomas L. Friedman. Farrar, Straus and Giroux, New York (1999)

112. Friedman, T.: The World is Flat: Brief History of the Twenty First Century, 660 p. Thomas L. Friedman. Farrar, Straus and Giroux, New York (2007)

113. Fukuyama, F.: America at the Crossroads: Democracy, Power, and the Neoconservative Legacy, 226 p. Francis Fukuyama. Yale University Press (2006)

114. Gamayunov, D.Y.: Detection of computer attacks based on the analysis of the behavior of network objects: dis. for the competition uch. degree of Cand. fiz.-mat. sciences. Moscow State University, Moscow (2007)

115. Gavrilova, T.A., Khoroshevsky, V.F.: Bases of Knowledge of Intellectual Systems: A Textbook for High Schools, 384 p. Peter, St. Petersburg (2000)

116. General report 2005. European Network and Information Security Agency. Brussels. 2005 [Electronic resource]. Access mode: https://www.enisa.europa.eu/publications/programmes-eports/enisa_work_programme_2005.pdf. Accessed 10 Apr 2016

117. General report 2008. European Network and Information Security Agency. 2009 [Electronic resource]. Access mode: https://www.enisa.europa.eu/publications/programmes-reports/enisa_gr_2008.pdf. Accessed date 10 Apr 2016

118. Global security in the digital age: stratagems for Russia; [under the Society. ed. AI Smirnova], 394 p. VNIIgeosistem, Moscow (2014)

119. Goldsmith, J.: Power and Constraint: The Accountable Presidency After 9/11, 311 p. Jack Goldsmith. W. W. Norton & Co., New York (2012)

120. GOST 15.000-94. System of product development and launching into manufacture. Basic provisions

121. GOST 22.0.05-97. Safety in emergencies. Technogenic emergencies. Terms and definitions

122. GOST 34.602. Information technology. Set of standards for automated systems. Technical directions for automated system making

123. GOST R 51583. Information protection. Sequence of automated operational system formation in protected mode. Basic provisions

124. GOST R 51624. Information protection. Protected automated systems. General requirements

125. GOST R MEK 61508–2012. Functional safety of electrical electronic programmable electronic safety-related systems. Part 1–7. Standartinform, Moscow (2014)

126. Graham, D.: Cyber Threats and the Law of War. David E. Graham. J. Natl. Security Law. 4(1), 87–102 (2010)

127. Grinyaev, S.N.: The battlefield – cyberspace: theory, methods, means, methods and systems of information warfare, 448 p. Harvest, Jordan (2004)

128. Guzik, V.F., Kalyaev, I.A., Levin, I.I.: Reconfigurable computing systems; [under the Society. ed. I.A. Kalyayeva], 472 p. Publishing house SFU, Rostov-on-Don (2016)

129. Hiller, J.: Internet Law & Policy. Janine S. Hiller, Ronnie Cohen. Prentice Hall, Upper Saddle River (2002). 377 p

130. Ilgun, K.: USTAT: A real-Time Intrusion Detection System for UNIX. Computer Science Department, University of California, Santa Barbara (1992)

131. Improving Critical Infrastructure Cybersecurity: Executive Order. The White House, Washington, DC. February 12, 2013 [Electronic resource]. Access mode: http://www.whitehouse.gov/the-press-office/2013/02/12/executive-orderimproving-critical-infrastructure-cybersecurity

132. Information Operations. Directive TS 3600.1. U. S. Department of Defense, Washington, DC. August 14, 2006 [Electronic resource]. Access mode: https://www.fas.org/irp/doddir/dod/info_ops.pdf

133. Information Operations. Joint Publication 3-13. Joint Chiefs of Staff, Washington, DC. November 27, 2012 [Electronic resource]. Access mode: http://www.dtic.mil/doctrine/new_pubs/jp3_13.pdf

134. Information Warfare. Directive TS 3600.1. U. S. Department of Defense, Washington, DC. December 21, 1992 [Electronic resource]. Access mode: http://www.dod.mil/pubs/foi/administration_and_Management/admin_matters/14-F-0492_doc_01_Directive_TS-3600-1. pdf

135. Intellectual Property and Development: Theory and Practice [ed. R.M. Olwan], 392 p. Springer, New York (2013)

136. International information security: problems and solutions; [under the Society. ed. S. A. Komov], 264 p. Moscow (2011)

137. International information security: world diplomacy: coll. materials; [under the Society. ed. S. A. Komov], 272 c. Moscow (2009)

138. International Strategy for Cyberspace. Prosperity, Security and Openness in a Networked World. The White House, Washington, DC. May 2011 [Electronic resource]. Access mode: http://www.whitehouse.gov/sites/default/files/rss_viewer/international_-strategy_for_cyberspace.pdf

139. Joint Doctrine for Information Operations. Joint Publication 3-13. Joint Chiefs of Staff, Washington, DC. October 9, 1998 [Electronic resource]. Access mode: http://www.c4i.org/jp3_13.pdf

140. Joint Doctrine for Information Operations. Joint Publication 3-13. Joint Chiefs of Staff, Washington, DC. February 13, 2006 [Electronic resource]. Access mode: http://www.bits. de/NRANEU/others/jp-doctrine/jp3_13(06).pdf

141. Joint statement by the Presidents of the Russian Federation and the United States of America on a new area of cooperation in building confidence. June 17, 2013 [Electronic resource]. Access mode: http://news.kremlin.ru/ref_notes/1479

142. Joint Terminology for Cyberspace Operations. Memorandum for Chiefs of the Military Services Commanders of the Combatant Commands Directors of the Joint Staff Directorates. The Vice Chairman of the Joint Chiefs of Staff, Washington, DC. [Electronic resource]. Access mode: http://www.nsciva.org/CyberReferenceLib/201011-Joint%20Terminology%20for%20Cyberspace%20Operations.pdf

143. Kalyaev, I.A., Levin, I.I., Semernikov, E.A., Shmoilov, V.I.: Reconfigurable Multicopy Computing Structures; [under the Society. ed. I. A. Kaliayev]. 2nd edn, 344 p. Pub. House of the Southern Scientific Center RAS, Rostov-on-Don (2009)

144. Kaplan, E.: Terrorists and the Internet. Council on Foreign Relations. January 8, 2009 [Electronic resource]. Access mode: http://www.cfr.org/terrorism-and-technology/terroristsinternet/p10005

145. Kaspersky, E.: Computer Malignity, 208 p. Peter, St. Petersburg (2008)

146. Kenneth, G.: Strategic Cyber Security. Copyright © 2011 by CCD COE Publications

147. Khaikin, S.: Neural Networks: Full Course = Neural Networks: A Comprehensive Foundation, 2nd edn, 1104 p. Williams, Moscow (2006)

148. Khomonenko, A.D., Tyrva, A.V., Bubnov, V.P.: Complex of programs for calculation of reliability and planning of software tests. Federal Service for Intellectual Property, Patents and Trademarks: Svid. about the state. reg. software for the computer ▢ 2010615617. Moscow (2010)

149. Khoroshevsky, V.G.: Architecture of Computing Systems. MSTU Them, 520 p. N.E. Bauman, Moscow (2008)

150. Kikot, S., Kontchakov, R., Podolskii, V., Zakharyaschev, M.: Query rewriting over shallow ontologies. Informal proceedings of DL 2013: 26th international workshop on description logics. CEUR workshop proceedings. vol. 1014, pp. 316–327 (2013)

151. Kikot, S., Tsarkov, D., Zakharyaschev, M., Zolin, E.: query answering via modal definability with FaCT++: first blood. Informal proceedings of DL 2013: 26th international workshop on description logics. CEUR workshop proceedings. vol. 1014, pp. 328–340 (2013)

152. Kikot, S., Zolin, E.: Modal definability of first-order formulas with free variables and query answering. J. Appl. Logic. **11**(2), 190–216 (2013)

153. Kilmburg, A.: Cybersecurity and Cyberpower: Concepts, Conditions and Capabilities for Cooperation for Action Within the EU. 2011 [Electronic resource]. Access mode: http://www.europarl.europa.eu/committees/fr/studiesdownload.html?languagedocment=en&file=41648

154. Kim, J., Bentley, P.: An Artificial Immune Model for Network Intrusion Detection. University College, London (1999)

155. Kissinger, H.: Does America Need a Foreign Policy? Toward a Diplomacy for the 21st Century, 238 p. Henry A. Kissinger. Simon & Schuster, New York (2001)

156. Klabukov, I.D., Alekhin, M.D., Musienko, S.V.: The sum of the national security and development technologies. Moscow (2014)

157. Klabukov, I.D., Alekhin, M.D., Nekhina, A.A.: The DARPA research program for 2015. Moscow (2014)

158. Kleschev, A.S., Artemieva, I.L.: Mathematical models of ontologies of subject domains. Part 2. Components of the model. STI. Ser. 2. **3**, 19–29 (2001)

159. Knake, R.: Cyberterrorism Hype v. Facts. Council on Foreign Relations. February 16, 2010 [Electronic resource]. Access mode: http://www.cfr.org/terrorism-andtechnology/cyberterrorism-hype-v-fact/p21434

160. Kohonen, T.: Self-Organizing Maps, 3rd edn. Springer, Berlin/New York (2001)

161. Kolmogorov, A.N.: Automats and life. In: Berg, A.I., Kolman, E. (eds.) Cybernetics: Expected and Cybernetics Unexpected, pp. 12–30. Science, Moscow (1968)

162. Korsakov, G.: Information weapons of the superpower. Ways. Peace. Secur. **1**(42), 34–60 (2012)

163. Kotenko, I.V.: Intellectual mechanisms of cybersecurity management. Proceedings of ISA RAS. Risk Manag. Safety, **41**, 74–103 (2009)

164. Kroes, N.: Speech: EU Cyber Security Strategy. Davos. 2013 [Electronic resource]. Access mode: http://europa.eu/rapid/press-release_SPEECH-13-51_en.html. Accessed date 10 Apr 2016

165. Krutskikh, A.V., Kramarenko, G.G.: Diplomacy and information and communication revolution. Int. Aff. **7**, 102–113 (2003)

166. Krutskikh, A.V.: To political and legal grounds of global information security. J. Int. Processes. [Electronic resource]. Access mode: http://www.intertrends.ru/thirteen/003.htm

167. Kumar, S., Spafford, E.H.: An Application of Pattern Matching in Intrusion Detection. Purdue University, New York (1994)

168. Kurnosov, M.G.: Models and algorithms for embedding parallel programs in distributed computing systems: Doctoral thesis in Technical. Science, 177 p. Siberian State University of Telecommunications and Informatics, Novosibirsk (2008)

169. Leibniz, G.V.: Essays in 4 volumes. Thought, Moscow (1982)

170. Letter dated 12 September 2011 from the Permanent Representatives of China, the Russian Federation, Tajikistan and Uzbekistan to the United Nations addressed to the Secretary-General. Document A/66/359 dated September 14, 2011 [Electronic resource]. Access mode: http://www.unmultimedia.org/radio/russian/wpcontent/uploads/2012/10/ga66359.pdf

171. Levakov, A.: A New priorities in the information security of the USA [Electronic resource]. Access mode: http://www.agentura.ru/equipment/psih/info/prioritet/

172. Levin, I.I., Dordopulo, A.I., Kalyaev, I.A., Doronchenko, Y.I., Razkladkin, M.K.: Modern and promising high-performance computing systems with reconfigurable architecture. Proceedings of the international scientific conference "Parallel Computing Technologies (PaVT'2015)", Ekaterinburg, March 31–April 2, 2015, pp. 188–199. Publishing Center of SUSU, Chelyabinsk (2015)

173. Levin, V.K., et al.: Communication network MVS-express. Inf. Technol. Comput. Syst. **1C**, 10–24 (2014)

174. Lewis, T.: Critical Infrastructure Protection in Homeland Security: Defending a Networked Nation, 474 p. Ted G. Lewis. Wiley-Interscience, Hoboken (2006)

175. Libicki, M.: Cyberdeterrence and Cyberwar. RAND Corporation. 214 p [Electronic resource]. Access mode: http://www.rand.org/content/dam/rand/pubs/monographs/2009/RAND_MG877.pdf (2009)
176. Libicki, M.: Cyberwar as a Confidence Game. Strat. Stud. Q. **5**(1), 132–146 (2011)
177. Libicki, M.: What Is Information Warfare? The Center for Advanced Command Concepts and Technology, 104 p. Institute for National Strategic Studies (1995)
178. Lisbon Summit Declaration. Issued by the Heads of State and Government participating in the meeting of the North Atlantic Council in Lisbon. November 20, 2010 [Electronic resource]. Access mode: http://www.nato.int/cps/en/natolive/official_texts_68828.htm#cyber
179. Lomov, B.F.: Methodological and Theoretical Problems of Psychology, 350 p. Science, Moscow (1999)
180. Lynn, W. III: Defending a New Domain: The Pentagon's Cyberstrategy. Foreign Affairs. September/October 2010 [Electronic resource]. Access mode: http://www.foreignaffairs.com/articles/66552/william-j-lynn-iii/defending-anew-domain
181. Lynn W. III: The Pentagon's Cyberstrategy, One Year Later. Foreign Affairs. September 28, 2011 [Electronic resource]. Access mode: http://www.foreignaffairs.com/articles/68305/william-j-lynn-iii/the-pentagonscyberstrategy-one-year-later
182. Malcolm, J.: Multi-stakeholder Governance and the Internet Governance Forum, 611 p. Terminus Press, Perth (2008)
183. Mamaev, M.A, Petrenko, S.A.: Technologies of Information Protection on the Internet, 848 p. Publishing house "Peter", St. Petersburg (2002)
184. Mansell, R.: Imagining the Internet: Communication, Innovation, and Governance, 289 p. Oxford University Press, Oxford (2012)
185. Markov, A.S., Tsirlov, V.L., Barabanov, A.V.: Methods for Assessing the Discrepancy Between Information Protection Means; [ed. A. S. Markov], 192 p. Radio and communication, Moscow (2012)
186. Markov, A.S.: Chronicles of cyberwar and the greatest redistribution of wealth in history. Quest. Cybersecurity. **1**(14), 68–74 (2016)
187. Marques J.F., Canessa N., Siri S., Catricala E., Cappa S. Conceptual knowledge in the brain: fMRI evidence for a featural organization Brain Res.. 2008. Vol. 1194. P. 90–99.
188. Marr, B.: Big Data: Using SMART Big Data, Analytics and Metrics to Make Better Decisions and Improve Performance, 246 c. Wiley, New York (2015)
189. Martin, A., Chao, L.L.: Semantic memory and the brain: structure and processes. Curr. Opin. Neurobiol. **11**, 194–201 (2001)
190. Massel, L.V.: Problems of smart grid creation in Russia from the perspective of information technologies and cyber security. In: Proceedings of the All-Russian Seminar with International Participation: Methodological Issues of Research into the Reliability of Large Energy Systems. Vol. 64. Reliability of energy systems: achievements, problems, prospects, pp. 171–181. ISEM SB RAS, Irkutsk (2014)
191. Masters, J.: Confronting the Cyber Threat. Council on Foreign Relations. May 23, 2011 [Electronic resource]. Access mode: http://www.cfr.org/technology-andforeign-policy/confronting-cyber-threat/p15577
192. Military Perspectives on Cyberpower [ed. L. Wentz, C. Barry, and S. Starr], 128 p. CreateSpace Independent Publishing Platform (2012)
193. Modern international relations and world politics: a textbook; [responsible. ed. A.V. Torkunov], 991 p. Education: MGIMO, Moscow (2004)
194. Moore, M.: Saving Globalization: Why Globalization and Democracy Offer the Best Hope for Progress, Peace and Development, 293 p. John Wiley & Sons (Asia), Hoboken (2009)
195. Mueller, M.: Networks and States: The Global Politics of Internet Governance, 313 p. Milton L. Mueller. MIT Press, Cambridge, MA (2010)
196. Multiple Futures: Navigating towards 2030. Final Report. NATO. Allied Command Transformation [Electronic resource]. Access mode: http://www.act.nato.int/images/stories/events/2009/mfp/20090503_MFP_finalrep.pdf

197. Murray, A.: Information Technology Law: The Law and Society, 602 p. Oxford University Press, Oxford (2013)
198. Nakashima, E.: Pentagon Proposes More Robust Role for its Cyber-specialists. The Washington Post. August 10, 2012 [Electronic resource]. Access mode: http://www.washingtonpost.com/world/national-security/pentagonproposesmore-robust-role-for-its-cyber-specialists/2012/08/09/1e3478ca-db15-11e1-9745-d9ae6098d493_story_1.html
199. Nathan, M., Warren, J.: Great data. Principles and practice of building scalable data processing systems in real time. Williams, Moscow (2016). 292 c. 90. Common internal security space in the EU: political aspects; [responsible. ed. S. V. Utkin], 146 p. IMEMO RAS, Moscow (2011)
200. National cyber security strategies. ENISA Report. 2012 [Electronic resource]. Access mode: http://www.enisa.europa.eu/activities/Resilienceand-CIIP/national-cyber-security-strategiesncsss/cybersecurity-strategies-paper
201. National Cybersecurity and Critical Infrastructure Protection Act of 2014. H.R. 3696 [Electronic resource]. Access mode: https://www.govtrack.us/congress/bills/113/hr3696
202. National Cybersecurity Strategy. Spain. 2013 2012 [Electronic resource]. Access mode: https://www.enisa.europa.eu/activities/Resilience-and-CIIP/national-cyber-security-strate gies-ncsss/NCSS_ESen.pdf. Accessed date 10 Apr 2016
203. National Military Strategy. A Strategy for Today; A Vision for Tomorrow. The Joint Chiefs of Staff, Washington, DC. 2004 [Electronic resource]. Access mode: http://www.defense.gov/news/mar2005/d20050318nms.pdf
204. National Security Strategy of the United States of America. The White House, Washington, DC. September 2002 [Electronic resource]. Access mode: http://www.state.gov/documents/organization/63562.pdf
205. National Strategy for the Physical Protection of Critical Infrastructures and Key Assets. The White House, Washington, DC. February 2003 [Electronic resource]. Access mode: http://www.dhs.gov/xlibrary/assets/Physical_Strategy.pdf
206. National Strategy for Trusted Identities in Cyberspace. Enhancing Online Choice, Efficiency, Security, and Privacy. The White House, Washington, DC, April 2011 [Electronic resource]. Access mode: http://www.whitehouse.gov/sites/default/files/rss_viewer/NSTICstrategy_041511.pdf
207. National Strategy to Secure Cyberspace. The White House, Washington, DC. February 2003 [Electronic resource]. Access mode: https://www.uscert.gov/sites/default/files/publications/cyberspace_strategy.pdf
208. NATO 2020: Assured Security; Dynamic engagement. Analysis and Recommendations of the Group of Experts on a New Strategic Concept for NATO. May 17, 2010 [Electronic resource]. Access mode: http://www.nato.int/cps/en/natolive/official_texts_63654.htm
209. Nye, J.: Cyber Power. Harvard Kennedy School. Belfer Center for Science and International Affairs. May 2010. 24 p. [Electronic resource]. Access mode: http://belfercenter.ksg.harvard.edu/files/cyber-power.pdf
210. Nye, J., Joseph, S.: Nuclear Lessons for Cyber Security? Strategic Studies Quarterly, 18–38 (2011)
211. On National and International Cyber Security Exercises: Survey, Analysis and Recommendations / European Network and Information Security Agency (ENISA). 2012 [Electronic resource]. Access mode: https://www.enisa.europa.eu/activities/Resilience-and-CIIP/cyber-crisis-cooperation/cce/cyber-exercises/exercise-survey2012
212. Order of the FSB of Russia and Federal Service for Technical and Export Control Order of Russia on August 31, 2010 No. 416/489 "On approval of the requirements for the protection of information contained in public information systems"
213. Ovdei, O.M., Proskudina, G.Y. Review of Ontology Engineering Tools, vols. 4–7. Institute of Software Systems, National Academy of Sciences of Ukraine, Kiev (2004)
214. Panarin, I.N., Panarina, L.G.: Information War and Peace, 384 p. OLMA-PRESS, Moscow (2003)

215. Pashchenko, I.N., Vasiliev, V.I., Guzairov, M.B.: Protecting Information in Smart Grid Networks Based on Intelligent Technologies: Designing the Rules Base, pp. 28–37. Izvestia YuFU. Technical science (2015)
216. Pashkov, V.: US information security. Foreign Military Rev. **10**, 3–13 (2010)
217. Petrenko, A.S., Petrenko, S.A.: Super-productive monitoring centers for security threats. Part 1. Protect. Inf. Inside. **2**(74), 29–36 (2017)
218. Petrenko, A.A., Petrenko, S.A.: Cyber units: methodical recommendations of ENISA. Quest. Cybersecurity. **3**(11), 2–14 (2015)
219. Petrenko, A.A., Petrenko, S.A.: Intranet Security Audit (Information Technologies for Engineers), 416 p. DMK Press, Moscow (2002)
220. Petrenko, A.A., Petrenko, S.A.: Research and Development Agency DARPA in the field of cybersecurity. Quest. Cybersecurity. **4**(12), 2–22 (2015)
221. Petrenko, A.A., Petrenko, S.A.: The way to increase the stability of LTE-network in the conditions of destructive cyber-attacks. Quest. Cybersecurity. **2**(10), 36–42 (2015)
222. Petrenko, A.S., Bugaev, I.A., Petrenko, S.A.: Master data management system SOPKA. Inf. Protect. Inside. **5**(71), 37–43 (2016)
223. Petrenko, A.S., Petrenko, S.A.: Designing of corporate segment SOPKA. Protect. Inf. Inside. **6** (72), 48–50 (2016)
224. Petrenko, A.S., Petrenko, S.A.: Large data technologies (BigData) in the field of information security. Inf. Protect. Inside. **4**(70), 82–88 (2016)
225. Petrenko, A.S., Petrenko, S.A.: Super-productive monitoring centers for security threats. Part 2. Protect. Inf. Inside. **3**(75), 48–57 (2017)
226. Petrenko, A.S., Petrenko, S.A.: The first interstate cyber-training of the CIS countries: "Cyber-Antiterror2016". Inf. Protect. Inside. **5**(71), 57–63 (2016)
227. Petrenko, S.A.: Methods of ensuring the stability of the functioning of cyber systems under conditions of destructive effects. Proceedings of the ISA RAS. Risk Manag. Security, **52**, 106–151 (2010)
228. Petrenko, S.A.: Methods of Information and Technical Impact on Cyber Systems and Possible Countermeasures. Proceedings of ISA RAS. Risk Manag. Security, **41**, 104–146 (2009)
229. Petrenko, S.A., Kurbatov, V.A., Bugaev, I.A., Petrenko, A.S.: Cognitive system of early warning about computer attack. Protect. Inf. Inside. **3**(69), 74–82 (2016)
230. Petrenko, S.A., Petrenko, A.A.: Ontology of cyber-security of self-healing SmartGrid. Protect. Inf. Inside. **2**(68), 12–24 (2016)
231. Petrenko, S.A., Petrenko, A.S.: Creation of a cognitive supercomputer for the computer attacks prevention. Protect Inf. Inside. **3**(75), 14–22 (2017)
232. Petrenko, S.A., Petrenko, A.S.: From detection to prevention: trends and prospects of development of situational centers in the Russian Federation. Intellect Technol. **1**(12), 68–71 (2017)
233. Petrenko, S.A., Petrenko, A.S.: Lecture 12. Perspective tasks of information security. Intelligent information radiophysical systems. Introductory lectures [A. O. Armyakov and others; ed. S.F. Boev, D.D. Stupin, A.A. Kochkarova], pp. 155–166. MSTU them. N.E. Bauman, Moscow (2016)
234. Petrenko, S.A., Petrenko, A.S.: New doctrine as an impulse for the development of domestic information security technologies. Intellect Technol. **2**(13), 70–75 (2017)
235. Petrenko, S.A., Petrenko, A.S.: New doctrine of information security of the Russian Federation. Inf. Protect. Inside. **1**(73), 33–39 (2017)
236. Petrenko, S.A., Petrenko, A.S.: Practice of application of GOST R IEC 61508. Inf. Protect. Insider. **2**(68), 42–49 (2016)
237. Petrenko, S.A., Shamsutdinov, T.I., Petrenko, A.S.: Scientific and technical problems of development of situational centers in the Russian Federation. Inf. Protect. Inside. **6**(72), 37–43 (2016)
238. Petrenko, S.A., Simonov, S.V.: Management of Information Risks. Economically Justified Safety (Information technology for engineers), 384 p. DMK-Press, Moscow (2004)

239. Petrenko, S.A.: The concept of maintaining the efficiency of cyber system in the context of information and technical impacts. Proceedings of the ISA RAS. Risk Manag. Safety. **41**, 175–193 (2009)
240. Petrenko, S.A.: The Cyber Threat model on innovation analytics DARPA. Trudy SPII RAN. **39**, 26–41 (2015)
241. Petrenko, S.A.: The problem of the stability of the functioning of cyber systems under the conditions of destructive effects. Proceedings of the ISA RAS. Risk Manag. Security. **52**, 68–105 (2010)
242. Petrenko, S.A., Asadullin, A.Y., Petrenko, A.S.: Evolution of the von Neumann architecture. Protect. Inf. Inside. **2**(74), 18–28 (2017)
243. Petrenko, S.A., Kurbatov, V.A.: Information Security Policies (Information Technologies for Engineers), 400 p. DMK Press, Moscow (2005)
244. Petrenko, S.A.: Methods of detecting intrusions and anomalies of the functioning of cyber system, Proceedings of ISA RAS. Risk Manag. Safety. **41**, 194–202 (2009)
245. Petrov, A.P.: On the perceptron's possibilities. Izvestiya AN SSSR, Technical Cybernetics. 6 (1964)
246. Portnoy, L., et al.: Intrusion detection with unlabeled data using clustering. ACM Workshop on Data Mining Applied to Security (2001)
247. Pospelov, D.A.: The modeling of reasoning. Experience in the analysis of mental acts, 184 p. Radio and communication, Moscow (1989)
248. Pospelov, D.A.: Thinking and Automatons, 224 p. 130. Soviet radio, Moscow (1972)
249. Presidential Decree of January 15, 2013 No. 31c "On the establishment of a state system for detecting, preventing and eliminating the consequences of computer attacks on Russia's information resources"
250. Primakov, E.M.: The world after September 11, 190 p. Thought, Moscow (2002)
251. Primakov, E.M.: Thoughts Aloud, 207 p. Rossiyskaya Gazeta, Moscow (2011)
252. Primakov, E.M.: A world without superpowers, in Russia. Global Polit. **3**, 80–85 (2003)
253. Protect Intellectual Property Act of 2011. S.968 [Electronic resource]. Access mode: https://www.govtrack.us/congress/bills/112/s968
254. Redko, V.G.: Evolution, Neural Networks, Intellect. LIBROKOM Book House/URSS, Moscow (2013)
255. Reich, V.V., Sinitsa, I.N., Sharashkin, S.M.: Model of the system for detecting attacks based on the detection of anomalies of network traffic. In: Proceedings of the second All-Russian Scientific Conference "Methods and Means of Information Processing"; [ed. member corr. RAS LN Koroleva], pp. 175–181. Izd. Department of the Factor of Computational Mathematics and Cybernetics of the Moscow State University. M.V. Lomonosov, Moscow (2005)
256. Report of governmental experts on achievements in the field of information and telecommunications in the context of international security. Document A/65/201 of 30 July 2010 [Electronic resource]. Access mode:http://www.un.org/disarmament/HomePage/ODAPublications/DisarmamentStudySeries/PDF/DSS_33_Russian.pdf
257. Report of the Group of Governmental Experts on Advances in Informatization and Telecommunications in the Context of International Security. Document A/68/98 of 23 June 2013 [Electronic resource]. Access mode: http://www.un.org/ga/search/view_doc.asp?symbol=A/68/98
258. Report on the implementation of the project for the implementation of the technological platform "Intellectual Energy System of Russia" (TPIS) in 2014 and the action plan for the TP IES for 2015 M. 2015. 93. Panarin I. Information war and power, 224 p. OLMA-PRESS, Moscow (2001)
259. Resolution expressing the sense of the Senate that the United Nations and other international organizations should not be allowed to exercise control over the Internet. S. RES. 323. November 18, 2005 [Electronic resource]. Access mode: https://www.govtrack.us/congress/bills/109/sres323/text

260. Right to privacy in the digital age. United Nations Resolution. Document A/RES/68/167 [Electronic resource]. Access mode: http://www.un.org/en/documents/ods.asp?m=A/RES/ 68/167

261. Rodin, A.V.: Logical and geometric atomism from Leibniz to Voevodsky. Prob. Philos. **6** (2016)

262. Rogov, S. M.: Doctrine of Bush. Free Thought. **XXI**(4), 4–4 (2002)

263. Rogov, S.M.: US Reaction and Consequences for Russian-American Relations, 88 p. ISKRAN, Moscow (2001)

264. Rogov, S.M.: USA at the Turn of the Century, 495 p. Science, Moscow (2000)

265. Rogov, S.M.: Obama's National Security Strategy, American Leadership in the Multipolar World, Independent Military Review. June 11, 2010. [Electronic resource]. Access mode: http://nvo.ng.ru/authors/4176/?PAGEN_1=2

266. Rogovskiy, E.A., American strategy of information predominance. Russia and America in the 21st century. Electronic J. **3** [Electronic resource]. Access mode: http://www.rusus.ru/? act=read&id=161%9A (2009)

267. Rogovskiy, E.A.: USA policy on securing cyberspace, USA, Canada. Econ. Polit. Culture. **6**, 3–22 (2012)

268. Rogovsky, E.A.: USA: Information Society. International relations, Moscow, 408 p (2008)

269. Rogowski, E.A.: Cyber-Washington: Global Ambitions, 848 p. International relations, Moscow (2014)

270. Roscini, M.: World wide warfare – Jus ad Bellum and the use of cyber force. In: von Bogdandy, A., Wofrum, R. (eds.) Max Planck Yearbook of United Nations Law, vol. 14, pp. 85–130 (2010)

271. Rosenzweig, P.: Cyber Warfare: How Conflicts in Cyberspace are Challenging America and Changing the World. Praeger, Santa Barbara (2013). 290 p

272. Russia-US bilateral project on cybersecurity. Fundamentals of critical terminology, 1st edn; [main. ed. Carl Frederick Rauscher, V. V. Yashchenko]. 2011 [Electronic resource]. Access mode: http://iisi.msu.ru/UserFiles/File/Terminology%20IISI%20EWI/Russia-U%20S%20% 20bilateral%20on%20terminology%20RUS.pdf

273. Ryzhikov, Y.I.: Work on the Thesis on Technical Sciences, 496 p. BHV-Petersburg, St. Petersburg (2005)

274. Sanger, D.: Confront and Conceal. Obama's Secret Wars and Surprising Use of American Power, 485 p. Broadway Paperbacks, New York (2012)

275. Sanger, D.: Obama order sped up wave of cyberattacks against Iran. The New York Times. June 1, 2012 [Electronic resource]. Access mode: http://www.nytimes.com/2012/06/01/world/ middleeast/obama-ordered-wave-ofcyberattacks-against-iran.html?pagewanted=all&_r=0

276. Sanger, D.: The Inheritance. A New President Confronts the World, 513 p. Black Swan (2009)

277. Schjolberg, S., Ghernaouti-Helie, S.: A Global Treaty on Cybersecurity and Cybercrime. Second edition. 89p. [Electronic resource]. Access mode: http://www.cybercrimelaw.net/ documents/A_Global_Treaty_on_Cybersecurity_and_Cybercrim,_Second_edition_2011.pdf (2011)

278. Schmidt, E., Cohen, J.: The New Digital World. How technologies change people's lives, business models and the notion of states; [trans. with English. S. Filin], 368 p. Mann, Ivanov and Ferber, Moscow (2013)

279. Schmitt, E.: Counterstrike: The Untold Story of America's Secret Campaign Against al Qaeda, 324 p. Eric Schmitt and Thom Shanker. Times Books, New York (2011)

280. Schmitt, M.: Wired Warfare: Computer Network Attack and Jus in Bello. International Committee of the Red Cross. RICR Juin IRRC. June 2002, Vol. 84, No 846, pp. 365–399. [Electronic resource]. Access mode: http://www.icrc.org/eng/assets/files/other/ 365_400_schmitt.pdf

281. Schneier, B.: The Eternal Value of Privacy. Wired. May 18, 2006 [Electronic resource]. Access mode: http://www.wired.com/politics/security/commentary/securitymatters/2006/05/70886. 333. Schwartau W. Cyber Shock: Surviving Hackers, Phreakers, Identity Thieves, Internet Terrorists, and Weapons of Mass Disruption, 470 p. Winn Schwartau. Thunder's Mouth Press, New York (2000)

282. Scott, D.S.: Models for various type-free calculi. Logic, Methodology and Philosophy of Science IV (Proc. Int. Congress 1971), pp. 157–188. North-Holland (1973)

283. Scott, D.S.: Outline of mathematical theory. 4th Annual Princeton Conference on Information Sciences and Systems, Princeton University, pp. 169–176 (1970)

284. Scott, D.S.: Logic and programming languages. Lectures of the winners of the Turing Award, pp. 65–83; [ed. R. Eschenhurst]. Mir, Moscow (1993)

285. Shakleina, T.A. Russia and the United States in World Politics: Textbook. Manual for University Students, 272 p. Aspect Press, Moscow (2012)

286. Shamir A., Wadge W.W. Data types as objects Lect. Notesin Corp. Sci. – 1977. No 52. P. 465-479.

287. Sharikov, P.A.: Approaches of Democrats and Republicans to the issues of information security. Russia and America in the 21st century. Electronic J. 1 (2012) [the Electronic resource]. Access mode: http://www.rusus.ru/?act=read&id=312

288. Sherstyuk, V.P.: Information security in the system of ensuring national security of Russia, federal and regional aspects of ensuring national security. Inf. Soc. 5, 3–5 (1999)

289. Sidnev, A.A., Gorshkov, A.V., Linev, A.V., Sysoev, A.V., Gergel, V.P., Kozinov, E.A., Meerov, I.B., Bastrakov, S.I.: Introduction to the principles of functioning and application of modern multinuclear architectures (by the example of Intel Xeon Phi). INTUIT, Moscow (2008) [Electronic resource]. Access mode: http://www.intuit.ru/goods_store/ebooks/9709/

290. Simonov, A.S., Slutskin, A.I., Leonova, A.E.: Directions of development of supercomputer technologies in JSC NICEVT. Inf. Technol. Comput. Syst. 2, 10–71 (2012)

291. Singer, P., Friedman, A.: Cybersecurity and Cyberwar: What Everyone Needs to Know, 306p. Oxford University Press, Oxford (2014)

292. Smelyansky, R.L.: Program-Configurable Networks, Open Systems. 5 (2012) [Electronic resource]. Access mode: http://www.osp.ru/os/2012/09/13032491/

293. Smirnov, A.A.: Providing information security in a virtualized society: the experience of the European Union. Monograph. UNITY-DANA, Moscow (2011). 196 p

294. Smirnov, A.I.: Global Security: Innovative Methods of Conflict Analysis. NIIGloB, Moscow. 272 (2011) with. [Electronic resource]. Access mode: http://niiglob.ru/index.php/en/20110115100852/18120110226201444.html

295. Smirnov, A.I.: Information globalization and Russia: challenges and opportunities, 392 p. NIIGloB, Moscow (2005). [Electronic resource]. Access mode: http://niiglob.ru/index.php/en/20110115100852/18020110226193238.html

296. Smirnov, A.I.: The fourth industrial revolution: information risks – a view from Russia. Int. Aff. Special Issue, 44–49 (2017)

297. Smirnov, A.I., Kokhtyulina, I.N.: Global security and "soft power 2.0": challenges and opportunities for Russia. VNIIgeosistem, Moscow (2012). 252 p. [Electronic resource]. Access   mode: http://niiglob.ru/index.php/en/20110115100852/307globalnayabezopasnostiqmyagkayasila20qvyzovyivozmozhnostidlyarossii.html

298. Smirnov, A.I.: Megatrends of Information Globalization. Yearbook of IMI. 3(13), 157–168 (2015)

299. Spinello, R.: Cyber Ethics: Morality and Law in Cyberspace, 238 p. Jones and Bartlett Publishers, Boston (2003)

300. Stankevich, L.A.: Artificial cognitive systems. Scientific session of National Research Nuclear University MEPhI -2010. XII All-Russian Scientific and Technical Conference "Neuroinformatics-2010". Lectures, pp. 106–160. National Research Nuclear University MEPhI, Moscow (2010)

301. Stop Online Piracy Act of 2011. H.R. 3261 [Electronic resource]. Access mode: https://www.govtrack.us/congress/bills/112/hr3261

302. Strategy to Combat Transnational Organized Crime: Addressing Converging Threats to National Security. The White House, Washington, DC. July 2011 [Electronic resource]. Access mode: http://www.whitehouse.gov/sites/default/files/Strategy_to_Combat_Transnational_Organized_Crime_July_2011.pdf

303. Streltsov, A.A.: Ensuring Information Security in Russia. Theoretical and methodological foundations; [ed. VA Sadovnichy and V. P. Sherstyuk]. Moscow Center For Continuous Mathematical Education, Moscow (2002). 296 p

304. Stupin, D.D., Kochkarov, A.A.: Organizational bases of pre-university preparation of students for high-tech companies of the economy real sector. Quality. Innov. Educ. **5**(72), 15–19 (2011)

305. Stupin, D.D., Kochkarov, A.A.: Prospects of the organization of pre-university youth training for high-tech companies in the real sector of the economy. Principles and mechanisms for the formation of the National Innovation System of the Russian Federation: coll. articles on the materials of the All-Russian scientific and practical conference, pp. 300–305. IE RAS, Moscow (2011)

306. Tallinn Manual on the International Law Applicable to Cyber Warfare. [Electronic resource] general editor Michael N. Schmitt. Cambridge University Press (2013). 282 p. Access mode: http://issuu.com/nato_ccd_coe/docs/tallinnmanual?mode=embed&layout=http%3A%2F%2Fskin.issuu.com%2Fv%2Flight%2Flayout.xml&showFlipBtn=true

307. Tarasov, A.M.: Electronic Government and Information Security: Manual, 648 p. GALART, St. Petersburg (2011)

308. Tarasov, V.B.: From Multiagent Systems to Intellectual Organizations (A series of "Sciences About the Artificial"), 352 p. Editorial URSS, Moscow (2002)

309. Tarasov, V.B.: System-organizational approach in artificial intelligence. Softw. Prod. Syst. **3**, 6–13 (1999)

310. Technology, Policy, Law and Ethics Regarding U. S. Acquisition and Use of Cyberattack Capabilities, 367 p. Ed by William Owens, Kenneth Dam, and Herbert Lin. National Academies Press, Washington, DC (2009)

311. Terekhov, V. A., Efimov, D.V., Tyukin, I.Y.: Neural Network Control Systems. Higher School, Moscow (2002). 184 p

312. Terrorist Use of the Internet: Information Operations in Cyberspace. Congressional Research Service. March 8, 2011. 16 p. [Electronic resource]. Access mode: http://www.fas.org/sgp/crs/terror/R41674.pdf

313. The concept of foreign policy of the Russian Federation (approved by the Decree of the President of the Russian Federation of November 30, 2016 No. 640

314. The concept of the development of an intelligent electric power system in Russia with an actively adaptive network. OJSC "FGC UES" OJSC "Scientific and technological center of electric power industry". Moscow (2011)

315. The concept of the state system for detecting, preventing and eliminating the consequences of computer attacks on the information resources of the Russian Federation (approved by the President of the Russian Federation on December 12, 2014, No. K 1274)

316. The Doctrine of Information Security of the Russian Federation (approved by the Decree of the President of the Russian Federation No. 646 of December 5, 2016).

317. The Economic Impact of Cybercrime and Cyber Espionage. The Center for Strategic and International Studies Report. July 2013. 19 p. [Electronic resource]. Access mode: http://csis.org/files/publication/60396rpt_cybercrimecost_0713_ph4_0.pdf

318. The European Cyber Security Month 2015: Deployment report. European Union Agency for Network and Information Security (ENISA). 2015 [Electronic resource]. Access mode: https://www.enisa.europa.eu/activities/stakeholder-relations/nis-brokerage-1/european-cyber-security-month-advocacy-campaign/2015. Accessed 10 Apr 2016

319. The national security strategy of the Russian Federation (approved by the Decree of the President of the Russian Federation of December 31, 2015, No. 683

320. The Order of the Ministry of Emergency Measures of the Russian Federation from February, 28th, 2003 ⊠ 105. On the statement of requirements on the prevention of extreme situations on potentially dangerous objects and objects of life-support

321. The Regulation on Cooperation of the Member States of the Collective Security Treaty Organization in the Sphere of Ensuring Information Security of December 10, 2010. [Electronic resource]. Access mode: http://docs.pravo.ru/document/view/16657605/14110649/. 129. Pospelov DA Thinking and automatons, 224 p. Soviet radio, Moscow (1972)

322. The role of science and technology in the context of international security, disarmament and other related fields. Report of the First Committee. Document A/53/576 of 18 November 1998 [Electronic resource]. Access mode: http://www.un.org/en/documents/ods.asp?m=A/53/576

323. The Russia U.S. Bilateral on Cybersecurity – Critical Terminology Foundations. EastWest Institute. Issue 1. April 2011. 47 p. [Electronic resource]. Access mode: http://www.ewi.info/idea/russia-us-bilateral-cybersecurity-criticalterminologyfoundations

324. The Stuxnet Computer Worm: Harbinger of an Emerging Warfare Capability. Congressional Research Service. December 9, 2010. 9 p. [Electronic resource]. Access mode: http://fas.org/sgp/crs/natsec/R41524.pdf

325. Thomas, T.: Cyber Silhouettes. Shadows Over Information Operations, 334 p. Timothy L. Thomas. Foreign Military Studies Office (FMSO). Fort Leavenworth (2005)

326. Thomas, T.: Is the IW paradigm outdated? A discussion of U.S. IW theory. J. Inf. Warfare. **2** (3), 109–116 (2003)

327. Threats Posed by the Internet. Threat Working Group of the CSIS Commission on Cybersecurity for the 44th Presidency. October 2008. 28 p. [Electronic resource]. Access mode: http://csis.org/files/media/csis/pubs/081028_threats_working_group.pdf

328. Toffler, A.: War and Anti-War: Survival at the Down of the Twenty-First Century, 1st edn, 302 p. Alvin and Heidi Toffler (1993)

329. Toffler, E.: The Third Wave, 784 p. AST, Moscow (2010)

330. Tsygichko, V.N., Votrin, D.S., Krutskikh, A.V., Smolyan, G.L., Chereshkin, D.S.: Information Weapons Are a New Challenge to International Security, 52 p. Institute of System Analysis of the Russian Academy of Sciences, Moscow (2000)

331. Tulving E. Episodic and Semantic Memory. Organization of Memory New York: Academic, 1972. P. 381–403.

332. Unsecured Economies: Protecting Vital Information. McAfee Report. (2009) 33 p. [Electronic resource]. Access mode: https://resources2.secureforms.mcafee.com/LP=2984

333. Vasyutin, S.V., Zavyalov, S.S.: Neural network method for analyzing the sequence of system calls for the detection of computer attacks and the classification of application modes. Methods and Means of Information Processing: Proceedings of the Second All-Russian Scientific Conference; [ed. member corr. RAS L.N. Koroleva], pp. 142–147. Pub. Department of the Factor of Computational Mathematics and Cybernetics of the Moscow State University. M.V. Lomonosov, Moscow (2005)

334. Velichkovsky, B.M.: Cognitive Technical Systems. Computers, Brain, Cognition: Successes of Cognitive Sciences, pp. 273–292. Nauka, Moscow (2008)

335. Vishnevsky, V.M., Lyakhov, A.I., Portnoy, S.L., Shakhnovich, I.V.: Broadband Wireless Information Transmission Networks. The technosphere, Moscow (2005)

336. Voevodin, V.V., Voevodin, V.L.B.: Parallel Computing, 609 p. BHV-Petersburg, St. Petersburg (2002)

337. Voevodsky, V.: Voevodsky V. A Very Short Note on the Homotopy Lambda-Calculus (2006)

338. Vorozhtsova, T.N.: Ontology as a basis for the development of an intellectual system for ensuring cybersecurity. Ontol. Des. **4**(14), 69–77 (2014)

339. Wales Summit Declaration. Issued by the Heads of State and Government participating in the meeting of the North Atlantic Council in Wales. September 5, 2014 [Electronic resource]. Access mode: http://www.nato.int/cps/en/natohq/official_texts_112964.htm

340. Weimann, G.: Cyberterrorism. How Real Is the Threat? United States Institute of Peace. Special Report. 12 p. [Electronic resource]. Access mode: http://www.usip.org/sites/default/files/sr119.pdf

341. Weimann, G.: Special Report 116: www.terror.net How Modern Terrorism Uses the Internet/ United Institute of Peace, March 2004. [Electronic resource]. Access mode: http://dspace.cigilibrary.org/jspui/bitstream/123456789/4610/1/www%20terror%20net%20How%20Modern%20Terrorism%20Uses %20the%20Internet.pdf?

342. Weimann, G.: Terror on the Internet: the New Arena, the New Challenges, 309 p. United States Institute of Peace Press, Washington, DC (2006)

343. Wesserman, F.: Neurocomputer Technology: Theory and Practice = Neural Computing. Theory and Practice, 240 p. Mir, Moscow (1992)

344. Widrow, B., Stirns, S.: Adaptive Signal Processing. Radio and communication, Moscow (1989)

345. Wiener, N.: Cybernetics, or Control and Communication in Animal and Machine. 2nd edn, 344 p. Science, Moscow; The main edition of publications for foreign countries (1983)

346. Wilshusen, G.: Cybersecurity: A Better Defined and Implemented National Strategy Is Needed to Address Persistent Challenges. Testimony Before the Committee on Commerce, Science, and Transportation and the Committee on Homeland Security and Governmental Affairs, U. S. Senate/United States Government Accountability Office. March 7, 2013. 36 p. [Electronic resource]. Access mode: http://www.gao.gov/assets/660/652817.pdf

347. Wilshusen, G.: Information Security: Cyber Threats and Vulnerabilities Place Federal Systems at Risk. Testimony Before the Subcommittee on Government Management, Organization, and Procurement; House Committee on Oversight and Government Reform/United States Government Accountability Office. May 5, 2009. 21 p. [Electronic resource].

348. Wilshusen, G.P., David, A.: Cybersecurity. Continued Efforts Are Needed to Protect Information Systems from Evolving Threats. Statement for the Record to the Subcommittee on Terrorism and Homeland Security, Committee on the Judiciary, U. S. Senate/United States Government Accountability Office. November 17, 2009. 24 p. [Electronic resource]. Access mode: http://www.gao.gov/new.items/d10230t.pdf

349. Wilson, C.: Botnets, Cybercrime, and Cyberterrorism: Vulnerabilities and Policy Issues for Congress/Congress Research Service Report. January 28, 2008. 40 p. [Electronic resource]. Access mode: http://fpc.state.gov/documents/organization/102643.pdf

350. Wilson, C.: Computer Attack and Cyber Terrorism: Vulnerabilities and Policy Issues for Congress/Congress Research Center Report. October 17, 2003. 32 p. [Electronic resource]. Access mode: http://fpc.state.gov/documents/organization/26009.pdf

351. Wilson, C.: Information Operations, Electronic Warfare and Cyberwar: Capabilities and Related Policy Issues/Congress Research Service Report. Updated March 20, 2007. 14 p. [Electronic resource]. Access mode: http://www.fas.org/sgp/crs/natsec/RL31787.pdf

352. Winterfeld, S.: The Basics of Cyber Warfare: Understanding the Fundamentals of Cyber Warfare in Theory and Practice, 164 p. Steve Winterfeld, Jason Andress. Syngress (2012)

353. Wolfengagen, V. E.: Categorical abstract machine. Lecture Notes: An Introduction to Computing. 2nd edn, 96p. JSC "Center YurInfo", Moscow. (2002)

354. Worldwide Threat Assessment of the US Intelligence Community for the Senate Select Committee on Intelligence: Office of the Director of National Intelligence. Statement for the Record. March 12, 2013. 34 p. [Electronic resource]. Access mode: http://www.intelligence.senate.gov/130312/clapper.pdf

355. Worldwide Threat Assessment of the US Intelligence Community for the Senate Select Committee on Intelligence: Office of the Director of National Intelligence. Statement for the Record. January 29, 2014. 31 p. [Electronic resource]. Access mode: http://www.intelligence.senate.gov/140129/clapper.pdf

356. Zakaria, F.: The Post-American World, 292 p. W. W. Norton, New York (2009)

357. Zdiruk, K.B., Astrakhov, A.V., Lonsky, A.V.: The model of information protection in heterogeneous computer networks based on the architecture of built-in "protected circuits". Proceedings of the Xth Russian Scientific and Technical Conference "New Information Technologies in Communication Systems and management", 1–2 June 2011, pp. 543–545. Kaluga (2011)

358. Zdiruk, K.B., et al.: Quest. Inf. Protect. **3**(78), 6–9 (2007)

359. Zhilyakova, L.Y.: The associative memory model based on a dynamic resource network. In: Proceedings of the conference "Management in technical, ergatic, organizational and network systems (UTEOSS2012)", pp. 1160–1163. State Scientific Center RF, JSC Concern CSRI Elektropribor, St. Petersburg (2012)

360. Zhukov, V.: The views of the US military leadership on the information warfare. Foreign Military Rev. **1**, 2–8 (2001)

361. Zinovieva, E.S.: International Internet Governance: Conflict and Cooperation: Textbook, 169 p. MGIMO-University (2011)